JTG Standards and Specifications

JTG Standards and Specifications

This series includes the standards and specifications of the highway transportation industry, issued by the Ministry of Transport of the People's Republic of China. It covers the standards and specifications in terms of technology, administration and service for the process from highway planning through to highway maintenance. The criteria for safety, environment and economic assessment are also included.

Yongshun Jiao

Hydrological Specifications for Survey and Design of Highway Engineering

人民交通出版社
China Communications Press

Springer

Yongshun Jiao
Hebei Provincial Communications Planning
and Design Institute
Beijing, China

ISSN 3004-8818 ISSN 3004-8826 (electronic)
JTG Standards and Specifications
ISBN 978-981-92-3127-0 ISBN 978-981-92-3128-7 (eBook)
https://doi.org/10.1007/978-981-92-3128-7

Jointly published with China Communications Press Co., Ltd
The print edition is not for sale in China (Mainland). Customers from China (Mainland) please order the print book from: China Communications Press Co., Ltd.
ISBN of the Co-Publisher's edition: 978-7-114-15928-2

This Springer imprint is published by the registered company Springer Nature Singapore Pte Ltd.
The registered company address is: 152 Beach Road, #21-01/04 Gateway East, Singapore 189721, Singapore

Foreword to Chinese Version

Following the requirements of *Notice on Planning of Revising Highway Engineering Technical Standard in 2011* (No. 115) issued by the Ministry of Transport of the People's Republic of China in 2011, the revision work for *Hydrological Specifications for Survey and Design of Highway Engineering* (JTG C30—2002) was carried out by Hebei Provincial Communications Planning and Design Institute.

In the revision process, the editorial team conducted extensive research, assimilated the research achievements and practical project design experience from the domestic scientific research institutes, universities, design institutes and so on, consulted and referred to the state-of-the-art international standards and specifications, as well as made comparisons and coordination with the relevant specifications in China. Considering the continuity of the revision work of the *Specifications*, Liu Xinsheng was invited to be the adviser of the revision work by the editing organization in charge.

The *Specifications* include 12 chapters and 5 appendices. The following contents are added in this edition.

1. According to the current relevant standards and specifications in combination with practical engineering experiences, one chapter about bridge site selection, which proposes the specifications of bridge site selection from the perspective for hydrological survey and design.
2. The content related to hydrological survey and design for reconstruction and expansion projects.
3. The content related to hydrological survey and design for river reaches with ice floe.
4. Adaptability analysis and evaluation of the current equations for calculating scour at piers and abutments were carried out to supplement and improve the application conditions of each equation.
5. One-dimensional riverbed sediment model was introduced for estimation of scour due to natural evolution of riverbed and general scour under bridge using numerical methods.

6. The content related to calculation equations for local scour of abutments in non-cohesive soil riverbed.
7. The content related to hydrological survey and design in gulf area.
8. The content related to the calculation for the spur dike protection length.

All relevant organizations are kindly requested to send and report problems and suggestions obtained from implementation of the *Specifications* to the management office for reference in the further revision. Please contact: Hua Pengnian (Address: Hebei Provincial Communications Planning and Design Institute, No.36, South Jianshe Street, Shijiazhuang, Hebei Province; Postal Code: 050011; Tel.: +86-311-86110931; E-mail: glswgfbxz@163.com).

Editing organization in charge for Chinese version:
Hebei Provincial Communications Planning and Design Institute

Associate editing organizations for Chinese version:
Sichuan Provincial Transport Department Highway Planning, Survey, Design and Research Institute
CCCC First Highway Consultants Co., Ltd.
Beijing Jiaotong University
Guangdong Provincial Highway Survey and Planning Design Institute, Ltd.

Chief editor for Chinese version: Jiao Yongshun

Associate editors for Chinese version:
Zhu Jijun
Hua Pengnian
Liang Linong
Qi Meilan
Song Guoping
Zhang Chunning
Zhang Liujun
Wen Fangzhen
Su Guanghe
Shen Peng

Chief review editor for Chinese version: Bao Weigang

Associate review editors for Chinese version:
Han Youxu
Shen Yonglin
Peng Li
Huang Meilan
Wang Lixin
Peng Baohua

Yang Yaoquan
Wang Sishun
Li Zhengrong
Peng Yuancheng
Li Huaifeng
Chen Zhen
Sheng Haifeng
Rong Guokai
Zhou Penghai
Han Dazhang
Zhang Qiuping
Lin Zhong
Jiang Xinting
Gao Dongguang

Beijing, China

The People's Republic of China, Ministry of Transport, Public Notice

No.36

Public Notice on Issuing the International Version of Seven Specifications of Highway Industry Standards

For promoting international cooperation in standardization of highway transportation industry, the Ministry of Transport hereby issues the international version of the following seven Specifications of Highway Industry Standards.

The French version of *Specifications for Highway Seismic Design*, JTG B02—2013(F);
The English version of *Hydrological Specifications for Survey and Design of Highway Engineering*, JTG C30—2015(E);
The English version of *Technical Guidelines for Design and Construction of Highway Embankment on Soft Ground*, JTG/T D31-02—2013(E);
The English version of *Design Specifications for Highway Drainage*, JTG/T D33—2012(E);
The French version of *Specifications for Design of Highway Asphalt Pavement*, JTG D50—2017(F);
The French version of *General Specifications for Design of Highway Bridges and Culverts*, JTG D60—2015(F); and
The English version of *Specifications for Design of Highway Tunnels—Section 2: Traffic Engineering and Affiliated Facilities*, JTG D70/2—2014(E).

The general administration and final interpretation of these Specifications belong to Ministry of Transport, while particular interpretation for application and routine administration of the international version of these Specifications shall be provided by China Road and Bridge Corporation.

In event of any ambiguity or discrepancies between the international version and Chinese version of these Specifications, the Chinese version should be referred and accepted.

Comments, suggestions and inquiries are welcome and should be addressed to China Road and Bridge Corporation by post address, by mail: 88C, Andingmenwai Dajie, Postal Code: 100011; or by E-mail: kjb@crbc.com.

It is hereby announced.

Ministry of Transport of the People's Republic of China
March 21, 2018

The People's Republic of China, Ministry of Transport, Public Notice

No.5

Public Notice On Issuing the *Hydrological Specifications for Survey and Design of Highway Engineering*

Hereby the *Hydrological Specifications for Survey and Design of Highway Engineering* (JTG C30—2015) is issued as one of the Highway Engineering Industry Standards, to become effective on March 1, 2015. The former edition of the *Hydrological Specifications for Survey and Design of Highway Engineering* (JTG C30—2002) and its English version, shall be superseded from the same date.

The general administration and final interpretation of the *Hydrological Specifications for Survey and Design of Highway Engineering* (JTG C30—2015) belong to Ministry of Transport, while particular interpretation for application and routine administration of the *Specifications* shall be provided by Hebei Provincial Communications Planning and Design Institute.

Comments, suggestions and inquiries are welcome and should be addressed to Hebei Provincial Communications Planning and Design Institute (Address: No. 36, South Jianshe Street, Shijiazhuang, Hebei Province; Postal Code: 050011). The feedbacks will be considered in future revisions.

It is hereby announced.

Ministry of Transport of the People's Republic of China
January 21, 2015

Introduction to English Version

Standards reflect the achievement of civilization, provide common languages for technical communications and improve global connectivity. In recent years, Chinese government has been proactively implementing the standardization to stimulate innovation, coordination, greening and opening up for shared development in China and worldwide. In light of mutual development along the Silk Road Economic Belt and the 21st Century Maritime Silk Road (so-called the Belt and Road Initiative), the Ministry of Transport of the People's Republic of China organized translation and published international version of Chinese transportation industry standards and specifications to cope with the increasing demands for international cooperation in world transportation, achieve interconnected development and promote knowledge dispersion and experience sharing.

JTG is the designation referring to the standards and specifications of highway transportation industry, issued by Ministry of Transport of the People's Republic of China. It covers the standards and specifications in terms of technology, administration and service for whole process from highway planning through to highway maintenance. The criteria for safety, environment and economy assessment are also included.

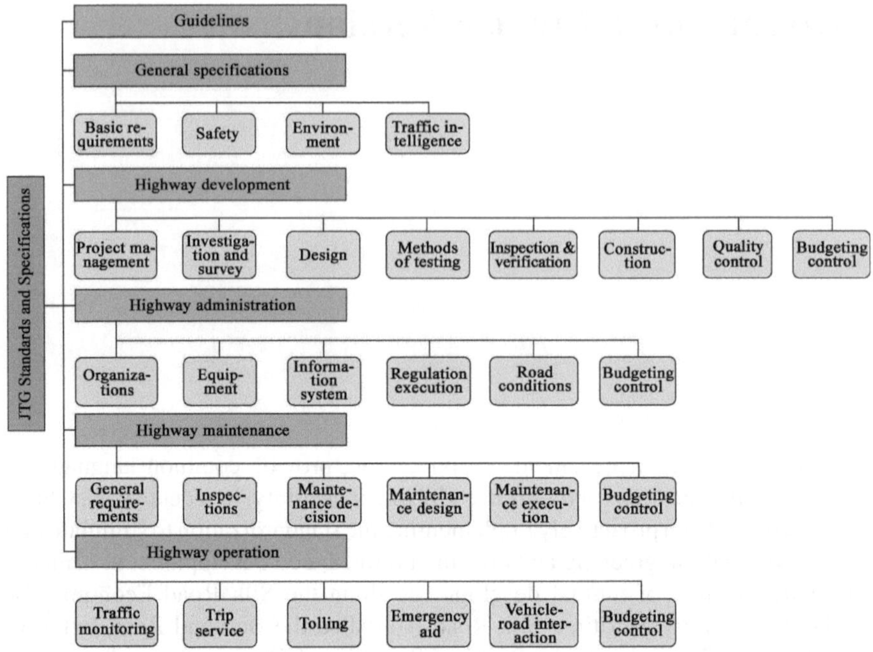

The original version of the *Hydrological Specifications for Survey and Design of Highway Engineering* (JTG C30—2015) is in Chinese language, which was updated in 2015 and taken into effective in March 1, 2015. The revision and updating of the English version was conducted by China Road and Bridge Corporation under the authorization of Ministry of Transport and Ministry of Housing and Urban-Rural Development of the People's Republic of China. It was approved by Highway Department of MoT China.

The contents and numbering of the chapters, sections, clauses and sub-clauses in the English version are exactly the same as those in Chinese version. In event of any ambiguity or discrepancies, the Chinese version should be referred and accepted.

Gratitude is given here to Mr. Hua Pengnian, one of the most important associate editors of the Chinese version for the valuable assistance and suggestions during the editing and approving of the English version.

Comments, suggestions and inquiries are welcome and should be addressed to editing organization in charge for English version (Address: 88C, Andingmenwai Dajie, Chaoyang District, Beijing, P.R. China; Postal Code: 100011; Fax: +86-10-64285907;E-mail: kjb@crbc.com). The feedbacks shall be taken into account in next editions.

Editing organization in charge for English version:
China Road and Bridge Corporation

Chief editors for English version:
Chen Baochun, Lu Shan

Associate editors for English version:
Fu Xiaolei, Jin Baoming

Review editor for English version:
Yao Haidong

Associate review editors for English version:
Chen Zhaohui, Wu Songjiang, Habib Tabatabai, Rosemarie Helmerich,
Khaled Sennah

Contents

Chapter 1
General Provisions

1.0.1 The *Specifications* are developed for regulating hydrological survey and design of highway engineering.

1.0.2 The *Specifications* are applicable to hydrological survey and design of new highways, upgrading or reconstructing of existing highways.

1.0.3 Hydrological survey and design of highway engineering cover hydrological investigation and survey of subgrade, bridges and culverts, hydrological and hydraulic analysis and computation, arrangement of bridges and culverts, and setting of regulation works, etc.

1.0.4 Hydrological investigation and survey shall be carried out through corresponding methods based on the design requirements and regional conditions of the project. Reliability assessment of the collected and investigated data shall be conducted, and the precision of the survey shall comply with the relevant provisions.

1.0.5 The rationality of hydrological and hydraulic analysis and computation results shall be verified. Hydrological tests and special demonstrations may be conducted for bridges with complicated hydrological conditions, with higher navigation level, and crossing sea.

1.0.6 Arrangement of bridges and culverts must meet the requirements for drainage, sediment transport and navigation, shall be compatible with road drainage system and water resource planning, and shall accommodate farmland irrigation and drainage.

1.0.7 Settings of water regulation works shall not affect the original function of the river reaches and the safety of river embankments (banks), villages and farmlands on both sides of the river.

1.0.8 Design flood frequencies for highway engineering structures shall comply with provisions specified in Table 1.0.8.

Y. Jiao, *Hydrological Specifications for Survey and Design of Highway Engineering*, JTG Standards and Specifications, https://doi.org/10.1007/978-981-92-3128-7_1

Table 1.0.8 Design flood frequencies

Structures	Highway class				
	Motorway	Class-1 highway	Class-2 highway	Class-3 highway	Class-4 highway
Super-large bridges	1/300	1/300	1/100	1/100	1/100
Large and medium bridges	1/100	1/100	1/100	1/50	1/50
Small bridges	1/100	1/100	1/50	1/25	1/25
Culverts and small drainage structures	1/100	1/100	1/50	1/25	Not required
Subgrade	1/100	1/100	1/50	1/25	Based on specific situations

Note 1. For super-large bridges on Class-2 highway and large bridges on Class-3 or Class-4 highway, in which the riverbed has a large gradient or is vulnerable to scour, the scour depth around bridge foundation should be checked with the design flood frequency increased by one level.

2. Design flood frequencies of longitudinal viaducts and bridge approaches along the river shall comply with the provisions for design flood frequency of this Table.
3. For a super-large bridge composed of multiple medium or small spans, its design flood frequency may adopt the criteria for large bridges.
4. Design flood frequency of the highway subgrade in the areas surrounding the cities shall be combined with the urban flood control standard, considering relief channel, flood drainage and discharge demand comprehensively

1.0.9 In addition to the *Specifications*, the hydrological survey and design of highway engineering shall also comply with the provisions in other current relevant national and industry standards.

Chapter 2
Terms and Definitions

2.0.1 Design flood.

The flood that meets the flood control design standard under normal engineering conditions.

2.0.2 Design flood frequency.

The statistical frequency of occurrence of the flood specified in accordance with the relevant technical standards as a design basis.

2.0.3 Design flood discharge.

Peak flood discharge at the bridge section corresponding to the design flood.

2.0.4 Design flood level.

The flood level corresponding to the design flood.

2.0.5 Structures in river affecting the bridge engineering.

All structures in river which affect the bridge work.

2.0.6 Backwater.

Water backed up upstream due to compressed water flow or push-up of water levels of tides or main streams.

2.0.7 Hydrological cross-section.

A cross section of a river that is selected for hydrological observation and hydrological analysis and computation.

2.0.8 Basic width of river channel.

The average width of channel formed by floods over many years, which is determined by the river hydraulic geometry.

Y. Jiao, *Hydrological Specifications for Survey and Design of Highway Engineering*, JTG Standards and Specifications, https://doi.org/10.1007/978-981-92-3128-7_2

2.0.9 Dominant discharge.

The discharge that controls the river formation and change.

2.0.10 The rising of local water surface in spate.

Local banked-up water level at the main stream of the section if the flood rises.

2.0.11 Roughness coefficient.

The coefficient that reflects the resistance of riverbed to flow.

2.0.12 Midstream of channel.

A line connecting the maximum velocity points at various cross sections along the river.

2.0.13 Thalweg.

A line connecting the maximum water depth points at various cross sections along the river.

2.0.14 Approach flow velocity.

The flow velocity at a certain distance to the upstream adjacent to a structure.

2.0.15 Threshold velocity.

Critical flow velocity at which stationary sediments on the riverbed begin to move.

2.0.16 Approach velocity during initial scour of pier.

The approach flow velocity in front of a pier at which the sediments at both sides of the pier upstream face begin to move under the shear stress of the streaming around the pier.

2.0.17 General scour.

Riverbed scour under a bridge caused by the increasing velocity of the flow compressed by the bridge span.

2.0.18 Local scour.

Scour of riverbed around piers and abutments caused by their obstruction to the water flow.

2.0.19 Safe value of headroom under bridge superstructure.

Safe value which is the sum of design water level plus various possible water level increment or that reserved above the maximum drift ice level.

2.0.20 Wave height.

The vertical height from the peak of wave to the trough of wave.

2.0.21 Wave run-up.

The vertical height at which the wave runs upward along a slope from the still water level.

2.0.22 Wave set-up.

Rising of water level formed at the upstream face of piers after the deformation of the waves if they run into the piers.

2.0.23 River hydraulic geometry.

The quantitative relationship among the riverbed morphology, the conditions for flow and sediment movement in the waters, and the composition of riverbed for a river at a state of dynamic equilibrium.

2.0.24 Drift-ice.

The phenomenon of water flow with floating ice or small pieces of ice.

2.0.25 Maximum velocity of flood and ebb.

Rapid flow velocities of flood and ebb, corresponding to the maximum flow velocity at the flood and ebb process, respectively.

Chapter 3
Scope and Requirements for Each Phase of Survey and Design

3.1 Feasibility Study Phase

3.1.1 The works of hydrological survey and design in the feasibility study phase shall provide hydrological information and reference for selecting route schemes, propose preliminarily the length of large and medium bridges, and estimate the amount of small bridges, culverts, protective works for submerged subgrade.

3.1.2 The basic data to be collected shall include the following contents:

1. The distribution of main rivers along the route, river catchment profile and river levee design standard, etc.
2. The class of navigable channel, channel map, the highest navigable level, the lowest navigable level, the minimum navigable height and the minimum navigable span.
3. The distribution of low-lying waterlogging area, flood-diversion area, flood detention (storage) area, and the operation of flood diversion and detention along the route.
4. The location, size, standard and operation of hydraulic project works along the route.
5. Information on water conservancy and river regulation planning.
6. The hydrological and meteorological information and flood calculation method in each area along the route.
7. Topographic maps with scales of 1:10,000 and 1:50,000.

3.1.3 Survey and analysis shall include the contents and requirements as follows:

1. The historical maximum flood level, normal water level, flood season, peak duration of general flood, drifting size, river frozen time and drift-ice, ice thickness, salivary drift-ice and so on, at the main river section where the bridge is located.

Y. Jiao, *Hydrological Specifications for Survey and Design of Highway Engineering*, JTG Standards and Specifications, https://doi.org/10.1007/978-981-92-3128-7_3

2. All previous flood diversion, bursting of each main river, the historical maximum water retention level, submerging time and water retention time at the flood diversion area, the flood detention (storage) area and the low-lying waterlogging area along the route.
3. River reach evolution near the bridge site of each main river, the visual geological composition of riverbed and measured width of river channel cross section.
4. Historical highest tide level, wave height along the route.
5. Shipping density of navigation channel, average maximum navigable water level over past years, minimum navigable water level, maximum ship tonnage, the scale of the fleet.
6. Draining width, height and usage of original bridge and culvert and all water conservancy projects along the route.
7. Measurement of the river gradient and the angle between the river and the route at the map in the scale of 1:10,000.
8. Design flood discharge for large or medium bridge may be calculated using the existing results or regional empirical equation.
9. Length of large or medium bridge over artificial rivers may be set by the design water surface width. Length of large and medium bridges over other rivers may refer to the effective width of water flow according to the historical maximum flood level, or set by the existing bridge length at upstream or downstream locations.
10. Span of small bridge and culvert may be set according to river cross-section width or existing span of small bridge and culvert.
11. The design of protection range of submerged subgrade.

3.1.4 Data compilation for hydrological survey and design shall include the following contents:

1. The explanation may include the following contents:

    (1) Date of investigation and survey.
    (2) Topographic type along the route.
    (3) Average annual precipitation over years, weather conditions and flood season.
    (4) Distribution of major long, medium rivers, low-lying waterlogging areas, flood diversion and detention (storage) areas along the route.
    (5) Flood and waterlog disasters over the years (submerging level and time).
    (6) Flood control standard of navigation channel and embankment projects.
    (7) Disaster conditions of river reaches with ice floe, the anti-collision facilities of the existing structure at the upstream and downstream locations, and their usage.
    (8) Situation and hydrogeology condition of the coast where the subgrade is located.
    (9) Planning of water conservancy and river regulation, suggestions for route setting, etc.

2. Results of hydrological investigation and survey in charts and tables.

3.2 Preliminary Design Phase

3.2.1 Hydrological survey and design in preliminary design phase shall include providing hydrological basis for route setting and design of subgrade, bridge and culvert works, etc., determining overall setting scheme of protective works for bridges and culverts and submerged subgrade.

3.2.2 Besides checking the basic data collected in the hydrological investigation and survey at feasibility study phase, the following data shall also be collected:

1. The hydrological data shall be collected at the preliminary survey according to the provisions in the current *Highway Survey Specifications* (JTG C10).
2. The relationship between the locations or elevations of horizontal control points and level points determined during preliminary survey or one-time location survey, and the elevations of level points set by relevant authorities.
3. The records of flood disaster in the count annals, local historical documents, newspapers and magazines, and the historical flood data collected by the local water conservancy and urban construction authorities.
4. Collecting the locations of upstream and downstream hydrology (water level) stations and tide stations, observed maximum flood discharge over years, the corresponding water level, the average sectional water velocity, the maximum water velocity, the gradient, the sand content, the observing date, the section of riverbed of the measuring station, the riverbed material, the elevation system, the correlation of the basic elements change in the flow measurement.

3.2.3 Survey and analysis calculations shall include the following contents and requirements:

1. Catchment profile investigation.
2. River morphology investigation and survey:
   (1) River investigation.
   (2) Hydrological cross-section mapping.
   (3) River gradient mapping.
   (4) Riverbed material measurement.
3. Investigation of flood and water level with other characteristics.
4. Flood observation.
5. Investigation of ice floe for river reaches with ice formation shall be conducted for large and medium bridges, and observation of ice floe over the river or reservoir with severe ice floe shall be conducted for super-large and large bridges.
6. Existing structures in river affecting the bridge engineering investigation. Super-large and large bridges with difficult technology under complex hydrology and

geology situations should conduct flood scour observation for the existing bridge at the bridge site.

7. Hydrological analysis and calculation:

    (1) Determining the design flood discharge, water level, flow direction and velocity.
    (2) Forecasting river evolution, sediment change near bridge site.
    (3) Drafting plane layouts of bridge (culvert) spans and regulation works, and predict forecast their influences on river reaches.
    (4) Determining the minimum height of bridges and culverts, and calculating the scour depth at piers, abutments and regulation works.

8. Hydrological survey and analysis calculation of small bridge and culvert, subgrade and special areas shall comply with the provisions of Chapters 9–11 in the *Specifications.*

3.2.4 Data compilation of hydrological survey and analysis calculation results shall include the following contents:

1. The explanation may include the following contents:

    (1) Date and scope of investigation and survey.
    (2) Topographic type, hydrological and weather conditions along the route.
    (3) Distribution and features of rivers, distribution and operation situations of main structures in river affecting the bridge engineering and its effect on discharge, discharge direction, flushing changes in bridge segments, water conservancy planning and scheme of river regulation.
    (4) Distribution and operation situations of low-lying waterlogging areas, flood diversion and detention (storage) areas, maximum flooding water level in history, flooding area, flood disaster and its mitigation.
    (5) Freeze-up and thaw periods, thaw situation and characteristics of various ice situations in river reach with ice floe.
    (6) Historical maximum tide level, heights of wave, ocean current, ice floe and sea surf, etc., of coastal subgrade.
    (7) Calculation results of design discharge, water level, velocity, scour depth at the foot of the subgrade slope along the river, and basis for arrangement of protection works.
    (8) River types in each long and medium bridge area, historical flood conditions, watercourse changes and prediction of its development tendency, safe flood discharge of watercourse, navigation requirements, floating objects and calculated results of design discharge, water level, velocity and scour depth, basis for determining bridge spans and setting regulation works.
    (9) Basis for calculating spans of small bridges and culverts, and reasons for arrangement of submersible bridges or overflow pavement.

(10) Design flood level of each low-lying waterlogging area, flood diversion and detention (storage) area, as well as the basis for arrangement of discharge structures.
(11) The change of hydrological condition after the reconstruction or expansion of existing bridges and culverts.
(12) Problems, to which attention shall be paid, in layout of route and design of subgrade, bridges and culverts.

2. Results of hydrological investigation and survey in charts and tables.
3. Results of design flood analysis and calculation in charts and tables.
4. Calculation of hydrological and hydraulic design.
5. Completing list of existing structures in the river affecting the bridge works.
6. Schematic drawings for general arrangement of bridges and culverts.
7. Schematic drawings for arrangement of subgrade regulation and protection works along the river and the coast.
8. Schematic drawings for arrangement of discharge constructions and protection works in low-lying waterlogging areas, flood diversion and detention (storage) areas.
9. Results of analysis of river pattern changes, argumentation of navigation, flood control evaluation and hydraulic model test.
10. Relevant letters and other data.

3.3 Construction Documents Design Phase

3.3.1 Hydrological survey and design in construction documents design phase shall comply with the following provisions:

1. Verifying the outcome data of hydrological investigation and survey at the preliminary design phase.
2. Testing and verifying the hydrological analysis and calculation results at the preliminary design phase, and spot checking the rationality of general arrangements of bridge (culvert) spans and settings of subgrade regulation and protection works.
3. Adding or revising the outcome data of hydrological investigation and survey and hydrological analysis and calculation results at the preliminary design to meet requirements for construction documents design.
4. Based on the review comments on the preliminary design, adding contents to the hydrological investigation and survey and adjusting results of hydrological analysis and calculation.

3.3.2 Hydrological survey and design for first phase of construction documents design shall be in accordance with Section 2 of the *Specifications*.

3.3.3 Contents of hydrological survey and design in this section may also be applicable to technical design phase.

Chapter 4
Bridge Site Selection

4.1 General

4.1.1 Bridge site selection shall conform to the route direction, except the controlled bridge site. Within the proper range along the river, bridge site selection may comprehensively take into account roads and the bridge according to the river reach characteristic of the hydrology, topography, geology, features, etc.

4.1.2 The bridge site for special large bridges with complex hydrology, geology and technology, shall be decided based on the determined general route direction, in addition, compared comprehensively on technology and economy in larger range along the river, according to river morphology, hydrology, geology, navigation requirement, ground facilities, construction condition and the relationship to the local economic and social development, etc. If necessary, geophysical exploration and drilling shall be carried out to ensure the implementability of bridge construction.

4.1.3 Bridge site selection in terms of hydrology shall comply with the following provisions:

1. Bridge site shall be selected at a straight, stable, and relatively narrow river reach.
2. Bridge site shall be selected by considering the natural evolution of the river, and the influence of the as-built bridge on the natural river.
3. Bridge axis should be perpendicular to the direction of the flow at medium and high flood level. If the bridge axis is skewed with the flow direction, its influence on hydrology shall be considered in the design of bridge span as well as foundations of piers and abutments.

4.1.4 Bridge site selection in navigable waters shall comply with the following provisions:

1. Bridge site shall be selected at the river reach with stable, straight navigation channel and with enough water depth for navigation. If the navigation channel is not stable, the influence caused by the change of the river way shall be considered.

Y. Jiao, *Hydrological Specifications for Survey and Design of Highway Engineering*, JTG Standards and Specifications, https://doi.org/10.1007/978-981-92-3128-7_4

2. Angle between a line normal to the bridge axis and the direction of navigation should not be larger than 5°. Otherwise, the navigation span shall be increased.
3. Bridge site shall be away from the existing hydraulic facilities, harbor working area and ship anchorage, etc.

4.1.5 As for bridges to be widened or reconstructed, if the existing bridges are located at a harbor area, an area with complex topography and features, or the area with curved or interlaced navigation channels, the bridge site may be re-selected. The distance between the proposed bridge location and the existing bridge location shall be designed by considering the requirements of navigation and flood control. The piers of the proposed bridge in water should be coordinated with those of the existing bridge.

4.2 Bridge Site Selection at Various Types of River Reach

4.2.1 For the river reach in mountain canyon with deep water and rapid flow, the bridge site should be selected at the location where one span can cross over.

4.2.2 For the river reach in open mountain area, the bridge site shall be selected at the location with stable river channel, relatively shallow water and slow flow.

4.2.3 For the river reach changed at the mountain front, the bridge site should be selected at the narrow river reach with relatively stable riverbanks and channel. If the bridge must cross over the diffused river, it shall be selected at the location where the channel swing range is small. Bridge axis should be perpendicular to the flood flow direction.

4.2.4 For the alluvial and sheetflood reaches at mountain front, the bridge site should be selected at the narrow reach of upstream or the contraction reach of downstream, and should not be selected at the diffused reach of middle stream.

4.2.5 For the straight or slightly curved river reach in plain, the bridge site should be selected at the location where the river channel has the same direction as the riverbed and the discharge of the river channel is large. The bridge axis should be perpendicular to the riverbank.

4.2.6 For the curved river reach in plain, the bridge site shall be selected at the relatively long river reach where the main channel has the same direction as the general river flow. If the river bend approaches to the basic riverside of the riverbed, the bridge site should be selected at the center of the top of the river bend.

4.2.7 For the branching river reach in plain, the bridge site should be selected at the upward of the thalweg branching point. For stable branching river reach at the islet in the river, bridge site may be selected at the point where the thalweg converges with the islet or its afterbody in the river.

4.2.8 For the river reach with wide floodplain, the bridge site should be selected at the river reach where the floodplain is at high elevation with small flow, and the channel is at the river center with stable and straight flow.

4.2.9 For the meandering river reach in plain, the bridge site should be selected at the long and narrow reach with fixed supports on the two riverbanks. The bridge axis should be perpendicular to the riverbank.

4.2.10 For the downdraft river reach, if the bridge crosses over the branch of the downdraft river, the site should be selected at the location where there is no or little influence by the inverse flow from the backwater of the main river.

4.2.11 For the tidal river reach, the bridge site should not be selected at the tidal bore section, shall avoid the unstable section at concave bank and shoal bank, and should not be in the vicinity of a tidal barrier.

4.2.12 For the river reach with ice-floe, the bridge site should be selected at the river reach with straight and stable channel, deep main channel, smooth ice flow, not at the reach with many floodplains, sand islets, branching and unsmooth water flow, where ice jam and ice dam are liable to occur.

4.3 Bridge Site Selection at Special Areas

4.3.1 Bridge site selection in reservoir area shall comply with the following provisions:

1. The change of river condition and the possible unfavorable effects caused by the construction of reservoir shall be considered.
2. Within area influenced by reservoir impoundment, bridge site should be selected at the location with narrow reservoir surface, stable bank slope and less sediment deposition. At frozen regions, it shall not be selected at the end of backflow or the location where ice dam is liable to occur.
3. At the downstream of reservoir, the bridge site should be selected outside the area influenced by downstream concentrated scour.

4.3.2 Bridge site selection in debris flow area shall comply with the following provisions:

1. It shall avoid the area where intense debris flow forms.
2. The bridge should neither be constructed by digging the channel, nor be connected by changing the channel.
3. If the route must pass through debris flow circulation area, the bridge site shall be selected at the straight reach in circulation area with stable gully bed, and shall be perpendicular to the main stream. It shall not be selected at the section where the gully bed grade changes from steep to flat and the cross-section contracts or expands suddenly, as well as at the turning point of bend.

4. For the route passing through the fan-shaped deposition of debris flow, the bridge site shall not pass through the middle and top area of the fan, and should be selected at the boundary and its of the fan. The bridge shall be separately set along the contour line. If the deposition fan close to the main river is subjected to the cutting by the water flow, the bridge site shall be selected by considering the development of the cutting and by setting a certain safe value.
5. For the route passing through the deposition fan group of debris flow, the bridge site should be selected at the exit of each gully or the end of transverse fan edge.

4.3.3 Bridge site selection in low-lying (river network) areas of plain shall comply with the following provisions:

1. Bridge site selection shall be matched with local planning of water conservancy and navigation, and should not be close to the water conservancy works such as water gate, the entrance of water drawing or flood diversion.
2. The bridge site should be selected at the location with high elevation on two banks, not at a section with silt or special and loose soil.
3. Wherever the bridge site crosses a network of irrigation channels, the existing irrigation system should not be destroyed.

4.3.4 Bridge site selection in karst region shall comply with the following provisions:

1. The strong karst regions should be avoided, and be selected in a region with slightly developed karst. If the bridge must be constructed at strong karst region, it shall be set at the location where the rock stratum is intact and the cave top is thicker.
2. The vast cavern, large shaft and structure fractured zone shall be avoided. If the structure fractured zone is not avoided, where shall be passed through by the bridge site by perpendicularly or with a small skew angle.
3. The bridge site should be set at the zone with indissoluble rock stratum, not at the contact zone between soluble and indissoluble rock strata.
4. Wherever the highway route crosses over the valley among peaks in karst hilly area, the bridge site should not be selected at the location like hopper, karst sinkhole, karst springs, underground river and its mouth.
5. The bridge site at the karst collapse regions shall be selected outside the hopper caused by the decrease of underground water level due to industry and civil water intake, and at the location with thicker covering, firm soil layer, and stable cave and underground water level.
6. The bridge should not be built within the region of underground river.

4.3.5 Bridge site selection in gulf region shall comply with the following provisions:

1. The bridge site should be selected at the location where the islands are connected and the flow cross-section is narrow.
2. The bridge site should be selected at the location where the bridge is connected smoothly to the highways of both banks, and the bridge axis is perpendicular to the ocean flow direction.

3. The bridge site should be selected at the location with stable coast, less sediment source and weak sediment flow along the coast. It should not be selected at the intersection with two or more sediment flows.
4. The bridge site selection should avoid ship anchorage.

Chapter 5
Hydrological Investigation and Survey

5.1 General

5.1.1 Hydrological investigation and survey shall provide basic data for hydrological analysis and calculation. The results of analysis and calculation shall be the basis to determine the design flood.

5.1.2 The main contents of hydrological investigation and survey shall meet the needs of hydrological analysis and calculation for projects, and comply with the relevant provisions in Chapter 3 of the *Specifications*.

5.2 Hydrological Investigation

5.2.1 The investigation of general situation of catchment areas should include the following contents:

1. Drawing the hydrographic chart along the route, checking the distribution of low-lying waterlogging areas, flood diversion area, flood detention (storage) areas, as well as locations and layouts of main water conservancy projects.
2. Measuring the characteristic values such as area, length, width, gradient of catchment area, as well as catchment area controlled by main water conservancy projects on topographic maps.
3. Investigating the distribution and extent of karsts, springs, debris flow, etc., and the characteristic data on soil types, topography, geomorphology and vegetation conditions.
4. Investigating water conservancy and river regulation planning information which may affect projects design in each catchment area.

5.2.2 Investigation of river reach should include the following contents:

Y. Jiao, *Hydrological Specifications for Survey and Design of Highway Engineering*, JTG Standards and Specifications, https://doi.org/10.1007/978-981-92-3128-7_5

1. Collecting graph papers and data of river reaches changing over the years, and investigating the development of river bend and conditions of floodplain and channel stability.
2. Investigating the phenomena such as tributaries, branches, rapids, detention spots, landslides, bank collapse and natural backwater.
3. Investigating flooding width and degree of stability of the river banks.
4. Investigating scour-silt changes of riverbed, upstream sediment sources, sedimentation height and cut depth in history.
5. Investigating the design standards of river embankment, safe discharge capacity of river channel and the corresponding water level.
6. Investigating the scheme of river regulation and its implementation time.
7. Investigating the class of navigable channel, the highest and lowest navigable water levels, number of navigable spans, locations of upward and downward navigation routes at high, medium, and low water levels, respectively.
8. Investigating the type and size of floating objects.
9. Determining flood roughness of each part of the riverbed according to riverbed shape, composition of sediment, bulkhead wall and vegetation conditions.

5.2.3 Flood investigation shall comply with the following provisions:

1. Based on collected data of historical flood, investigating the occurrence time of floods, flood-mark spot, flood sources, fluctuation processes and main stream direction of each flood on both banks, investigating whether there is overland flow, branches and effects of artificial structures, determining return period of flood.
2. Situations of rainfall and flood, information about whether human activities are affecting the catchment area, and the change of natural conditions of each flood shall be investigated; determining the return period of flood according to their previous scale.
3. The river reach where flood is investigated should be close to the hydrological section where there are many flood marks on both sides of the bank, the stream is straight and steady, and there is no back water or flood diversion or effects of artificial constructions.
4. Three or more reliable flood marks shall be investigated for one flood. It is necessary to make marks, and record the name, occupation, age, and description of the person designating the flood marks. According to the provisions in Table 5.2.3, the reliability of flood marks may be analyzed and judged comprehensively based on the conditions of specified flood landmarks and the recollection of witnesses regarding the flood.

5.2.4 In the flood investigation, low water level, ordinary water level, and water surface cross slope, rising of local water surface in spate and wave height of flood period shall be investigated.

5.2.5 Ice floe investigation should include the following contents:

1. Investigating freeze-up and thaw periods over the years, breaking-up situation, highest and lowest level of floating ice.

Table 5.2.3 Evaluation standards of flood marks reliability

Evaluation factors	Degree		
	1	2	3
	Reliable	Less reliable	Reference
Impression of witnesses, the circumstantial evidence	Witnessed personally, deep impression, lifelike circumstance, and definite other evidence	Witnessed personally, deep impression, less lifelike circumstance, few other evidence	Heard from someone else or witness less certain, not a clear and specific circumstance, no evidence
Landmarks and flood mark situations	Fixed landmarks, specific flood marks site or clear flood marks	Landmarks nearly unchanged, specific flood marks site	Landmarks changed, no specific flood marks site
Estimating the range of reliability error (m)	<0.2	0.2–0.5	0.5–1.0

Note Degree 1 and 2 are mainly used to evaluate the reliability, Degree 3 is only for reference. They shall be used according to specific circumstance

2. Investigating ice jam and ice dam phenomenon, historical damage of ice run and the impact of floating ice to upstream and downstream structures.

5.2.6 Investigation of structures in river affecting the bridge engineering should include the following contents:

1. Design flood standards of any existing bridges, spans of pipes and cables crossing over the river, embedded depth of the foundation, age of construction, water damages and protections, etc. of river reach near bridge site.
2. Design flood standards of dam, structure form, embedded depth of foundation, construction quality and flood inspection.
3. Reservoir sites in upstream and downstream, design flood standards, flood discharge rates, controlled catchment areas, backwater boundaries, scour-silt changes of riverbed in upstream and downstream after the construction of reservoir.
4. Locations of structures in river affecting the bridge projects, such as water intake, pumping stations, wharfs, timber yards, anchorage ground, etc., and its influence on highway engineering.

5.3 Hydrological Survey

5.3.1 Surveying and mapping of hydrological cross-section shall comply with the following provisions:

1. Hydrological cross-section should be selected at a location where there are more flood-marks, stable bank, little scour-silt, narrower spreading width, no stagnant water or backwater, straight river reach with regular cross sections which should be perpendicular to the stream.
2. Hydrological cross-section shall be surveyed and mapped in both upstream and downstream of the bridge site, and may be surveyed and mapped of a medium bridge where the river is not wide; if conditions of bridge site section meets the requirements of hydrological cross-section, it may be used as hydrological cross-section. For reconstruction or expansion projects, the bridge hydrological cross-section shall be mapped, and one or two hydrological cross-sections shall be selected for mapping at the river reach where it is not affected by the bridge.
3. Range of surveying and mapping in rivers with wide floodplain in plain areas shall cover 50 m beyond the historically highest flooding line; and over the historical highest flood level by 2–5 m for rivers in mountainous areas.
4. Riverbed ground line, boundaries between floodplain and channel, vegetation and geological conditions, roughness, sounding level, measuring time, historical flood level and its outbreak year, other characteristic level of water, etc. shall be marked. Boundaries between floodplain and channel shall be determined on site.

5.3.2 Surveying and mapping for gradient of river reach shall comply with the following provisions:

1. The range of hydrological cross-section for surveying and mapping shall not be less than the river width in downstream areas, and not be less than twice the river width in upstream areas.
2. Riverbed gradient line, water surface gradient line during measurement, gradient line of previous floods, locations of hydrological cross-section and bridge site section shall be marked.

5.3.3 Test of riverbed material shall comply with the following provisions:

1. Test of riverbed material shall be according to geological exploration data to determine the type, properties and average particle size of riverbed materials in each layer of riverbed section.
2. Surface riverbed materials may be determined according to the current *Geotechnical Test Regulations for Highways* (JTG E40) through collecting disturbed soil samples, conducting particle analysis and performing liquid and plastic limit experiments.
3. The number of soil samples collected in river channel shall not be less than 1 for small bridges, not less than 2 for medium bridges and not less than 3 for large and super-large bridges; 1–2 soil samples may be collected in floodplain areas depending on the soil distribution.
4. Depth of samples shall be larger than the thickness of sediment transport.

5.3.4 Ice floe observation shall comply with the following provisions:

1. If ice thaws in spring, the following should be observed: river central ice thickness, ice temperature, ice block size, velocity and direction, ice layer area, length along the flow direction, water velocity under ice layer, water surface gradient, wind speed and direction, temperature change rate and other data that are necessary for ice pressure calculation.
2. Observation period should not be less than one freezing period, and observations should be conducted every 5 days, or observations shall be conducted every 1–2 days if necessary.

5.3.5 In addition to conventional hydrological survey for bridge reconstruction or expansion projects, the relationship between existing structures and structures to be constructed, and the ranges of the existing flood disaster and the repair works shall be surveyed and mapped, and the scour depth at piers and abutments of existing bridges shall be measured.

5.4 Flood Observation

5.4.1 Flood observation shall be conducted for complex hydrological conditions or special bridges that need hydraulic model testing. The observation items may be determined if necessary, and should include water level, water depth, velocity, direction, hydrological cross-section, water surface gradient and sediment content. For general bridge survey under flooding conditions, observation of water level, velocity, flow direction, gradient, etc., should be conducted, and hydrological cross-section is observed after flood.

5.4.2 Water level and surface gradient observation shall comply with the following provisions:

1. Basic water gauge or self-recording level gauge shall be set at the hydrological cross-section to observe water level; slope gauges shall be set in upstream and downstream of the basic gauge to observe water surface gradient; basic gauge coupled with slope gauge in upstream or downstream may also be used. Spacing between slope gauges should not be less than the required distances shown in Table 5.4.2.
2. Observation period and precision of basic water gauge should be consistent to upstream and downstream hydrological (level) stations. In flood periods if water level changes abruptly, observation frequency shall be increased and peak flood process shall be controlled.
3. Water levels at upstream and downstream gradient gauges should be observed simultaneously, and their observation frequency shall be consistent with the frequency of flow velocity observation.
4. Flood level observation may be carried out once during flood rising, peak flood and flood falling, respectively, and meanwhile water surface gradient is observed along upstream and downstream of hydrological cross-sections. The water level

Table 5.4.2 Spacing between water level gauges

Gradient (%)	0.068	0.038	0.028	0.022	0.019	0.016	0.015	0.013	0.012	0.011
Spacing between water gauges (m)	100	200	300	400	500	600	700	800	900	1000

difference along the total length of observation should be controlled within 0.1–0.3 m, of which the length along the upstream accounts for 2/3 and 1/3 along the downstream.

5.4.3 Flow velocity observation shall comply with the following provisions:

1. Current meters should be used for measuring flow velocity. If this is difficult, even buoy method may be adopted. If the measuring time needs to be shortened due to the short duration of flood peak, the midstream buoy method may be used.
2. Velocity observations shall be conducted during at least one flood peak event. For each flood peak event, at least twice observations shall be made before the peak flood, once near the peak flood and twice after the peak flood. Meanwhile, the water level, wind speed and wind direction shall be observed.
3. For general bridges, the method with a midstream buoy or a floater buoy may be used to monitor the velocity, and each observation shall be conducted before and after the flood peak and near the flood, respectively.

5.4.4 Flow direction observation should comply with the following provisions:

1. Flow direction vanes, flow direction apparatus and buoys may be used for monitoring flow direction. If buoys are used, flow direction observation should be conducted at the same time as the velocity observation.
2. If buoys are used for measuring the length of river reach, the length along the upstream of a hydrological section or the bridge site should be no less than twice the river width and no less than the river width along the downstream of the hydrological section or the bridge site.
3. Flow direction should be determined by tracking movements of buoys.

5.4.5 Water depth and hydrological section should be measured simultaneously with velocity measurement, and shall comply with the relevant provisions in the Clause 5.3.1 of the *Specifications*.

Chapter 6
Analysis and Calculation of Design Flood

6.1 General

6.1.1 Design flood shall correspond with the annual highest flood discharge and the discharge hydrograph associated with the specified frequency indicated in Clause 1.0.8 of the *Specifications*. If the structure is designed by water level control, the design of flood level shall comply with the annual highest flood discharge with the specified frequency shown in Clause 1.0.8 of the *Specifications*. If design flow is calculated according to storm runoff, the rainfall intensity or precipitation shall comply with the assigned frequency in Clause 1.0.8 of the *Specifications*.

6.1.2 The reliability, consistence and series representativeness of the flood data shall be examined for analysis and calculation.

6.1.3 Flood analysis and calculation may be carried out using several methods according to the details of data and the regional characteristics. The reasonable analysis and calculation results may be adopted after analysis and verification.

6.2 Calculation of Design Flood Discharge by Measured Discharge Data

6.2.1 Examination and selection of measured discharge data shall comply with the following provisions:

1. The maximum flood discharge measured in each year shall be selected with the identical flood type and conforming to the independent and random conditions.
2. The maximum flood discharge measured in each year shall be modified and reconstructed according to natural conditions, if there are artificial influence, natural channel bursting or diversion, etc.

Y. Jiao, *Hydrological Specifications for Survey and Design of Highway Engineering*,
JTG Standards and Specifications, https://doi.org/10.1007/978-981-92-3128-7_6

3. The maximum flood discharge measured in different periods shall be revised based on the correlation of previous variations if there are changes to basic elements such as station site or benchmark plane.
4. For some primary items from the measured flood discharge data series, their reliability shall be examined through the analysis and comparison of basin flood or the field investigation.
5. If flood frequency is calculated, the measured flood discharge data series should not be less than 30 years, and historical flood investigation and research results shall be available.

6.2.2 Inserting, extending and converting data series of measured flood discharge shall comply with the following provisions:

1. If the difference between catchment areas in hydrological calculation cross-section and hydrological station is less than 20% of catchment area of hydrological station and no more than 1000 km^2, and storm distribution is comparatively even in catchment area within which there are no flood diversion and detention areas, the measured maximum flood discharge at hydrological station may be converted into flood discharge at hydrological calculation cross-section by the following equation:

$$Q_1 = \left(\frac{F_1}{F_2}\right)^{n_1} Q_2 \tag{6.2.2}$$

where:

Q_1, F_1 flood discharge (m^3/s) and catchment area (km^2) at hydrological calculation cross-section, respectively;

Q_2, F_2 measured maximum flood discharge (m^3/s) and catchment area (km^2) at hydrological station, respectively;

n_1 a regional empirically-determined value, generally $n_1 = 0.5$–0.7 for long and medium rivers, $n_1 \geq 0.7$ for small rivers with catchment area smaller than 100 km^2.

2. If the measured flood level series is longer than measured flood discharge series, or if flood discharge is not measured while data of measured flood level is available, relationship curves of measured water level versus discharge should be established to insert or extend for flood discharge data series.
3. The number of years used in inserting or extending should not be more than the years of measured flood discharge, and rationality analysis shall be conducted combining meteorological and geographical conditions.

6.2.3 Calculation of empirical frequency of flood discharge shall comply with the following provisions:

1. For continuous series, empirical frequency may be estimated by the following equation:

$$P_m = \frac{m_i}{n+1} \times 100 \tag{6.2.3-1}$$

where:

P_m empirical frequency of measured flood discharge (%);
m_i sequence number based on measured flood discharge series in descending order;
n total number of measured flood discharge series.

2. For discontinuous series, one of the following methods may be used to estimate empirical frequency:

(1) Catastrophic flood discharge and measured flood discharge investigated in N years during the investigating period are ranked in their respective data series; the empirical frequency of the measured flood discharge may be estimated by Eq. (6.2.3-1), and that of the catastrophic flood discharge may be estimated by the following equation:

$$P_M = \frac{M}{N+1} \times 100 \tag{6.2.3-2}$$

where:

P_M empirical frequency of historical catastrophic flood discharge or catastrophic flood discharge in measured flow series (%);
M sequence number of historical catastrophic flood discharge or catastrophic flood discharge in measured flow series during the survey period;
N number of years for investigation period.

(2) A discontinuous series is composed of measured flood discharge and catastrophic flood discharge investigated in N years during the survey period. Empirical frequency of catastrophic flood discharge may be estimated by Eq. (6.2.3-2), and that of other measured flood discharges may be estimated by the following equation:

$$P_m = \left[\frac{a}{N+1} + \left(1 - \frac{a}{N+1}\right)\frac{m_i - l}{n-l+1}\right] \times 100 \tag{6.2.3-3}$$

where:

P_m empirical frequency of measured flood discharge (%);
a item number of catastrophic flood;
l item number of terms in measured flood discharge series which are processed as catastrophic flood discharge.

6.2.4 The Pearson III curve should be adopted as the theoretical frequency curve; however, other curves may also be used in special circumstances after analysis and verification.

6.2.5 To calculate the statistical parameters of the frequency curve, moment-based curve-fitting method, three-point curve-fitting method and modified moment method based on line-plotting and value-peaking may be used to calculate preliminary calculation value of mean value $\overline{Q}$, deviation coefficient C_V and coefficient of skew C_S of flood discharge series. A plot of the theoretical frequency curve is compared with the empirical frequency data of measured flow. If they do not match well, C_V and C_S may be adjusted until they coincide with each other for the most part.

6.2.6 Design flood discharge shall be calculated by the following equation based on the adjusted frequency curve parameters:

$$Q_p = \overline{Q}(1 + \Phi_p C_V) \tag{6.2.6}$$

where:

Q_p design flood discharge (m^3/s);
Φ_p coefficient of deviation from the average.

6.3 Calculation of Design Flood Discharge by Historical Flood Level

6.3.1 Calculation of historical flood discharge may use one of the following methods:

1. If the surveyed historical flood level in river reaches with steady and uniform flow, where the water surface gradient is uniform, the river channel is straight and the cross section of riverbed is relatively regular, the historical flood discharge may be calculated by the following equations:

$$Q = A_c v_c + A_t v_t \tag{6.3.1-1}$$

$$v_c = \frac{1}{n_c} R_c^{\frac{2}{3}} I^{\frac{1}{2}} \tag{6.3.1-2}$$

$$v_t = \frac{1}{n_t} R_t^{\frac{2}{3}} I^{\frac{1}{2}} \tag{6.3.1-3}$$

where:

Q historical flood discharge (m^3/s);
A_c, A_t discharge area of channel and floodplain respectively (m^2);
v_c, v_t mean velocity of channel and floodplain respectively (m/s);
n_c, n_t roughness of channel and floodplain respectively;
R_c, R_t hydraulic radius (m) of channel and floodplain respectively, which may be replaced by average water depth if breadth-to-depth ratio is larger than 10;
I water surface gradient.

2. If the surveyed historical flood level in river reaches with steady but non-uniform flow, where there are significant difference of shape and area of riverbed cross section, the historical flood discharge may be calculated by the following equations:

$$Q = \overline{K}\sqrt{\frac{\Delta H}{L\text{-}\left(\frac{1\text{-}\xi}{2g}\right)\left(\frac{\overline{K}^2}{A_1^2}\text{-}\frac{\overline{K}^2}{A_2^2}\right)}} \quad (6.3.1\text{-}4)$$

$$\Delta H = H_1 - H_2 \quad (6.3.1\text{-}5)$$

$$\overline{K} = \frac{1}{2}(K_1 + K_2) \quad (6.3.1\text{-}6)$$

$$K_1 = \frac{1}{n_{c1}}A_{c1}R_{c1}^{\frac{2}{3}} + \frac{1}{n_{t1}}A_{t1}R_{t1}^{\frac{2}{3}} \quad (6.3.1\text{-}7)$$

$$K_2 = \frac{1}{n_{c2}}A_{c2}R_{c2}^{\frac{2}{3}} + \frac{1}{n_{t2}}A_{t2}R_{t2}^{\frac{2}{3}} \quad (6.3.1\text{-}8)$$

where:

H_1, H_2 water level at upstream and downstream sections respectively (m);
ΔH water head of upstream and downstream sections (m);
L distance between upstream and downstream sections (m);
A_1, A_2 total discharge area of upstream and downstream sections, respectively (m^2);
A_{c1}, A_{t1} discharge area of channel and floodplain, respectively, at upstream section (m^2);
A_{c2}, A_{t2} discharge area of channel and floodplain, respectively, at downstream section (m^2);
R_{c1}, R_{t1} hydraulic radius of channel and floodplain, respectively, at upstream section (m);
R_{c2}, R_{t2} hydraulic radius of channel and floodplain, respectively, at downstream section (m);
n_{c1}, n_{t1} roughness of channel and floodplain, respectively, at upstream section;
n_{c2}, n_{t2} roughness of channel and floodplain, respectively, at downstream section;
K_1, K_2 water delivery coefficient of upstream and downstream sections, respectively (m^3/s);
$\overline{K}$ average value of discharge modification coefficients of upstream and downstream sections (m^3/s);
g 9.80 (m/s^2);
ξ coefficient of local head loss. If it reduces towards downstream, a value of −0.1-0 is taken; if it increases gradually towards downstream, a value of 0.3–0.5 is taken; if it increases abruptly towards downstream, a value of 0.5–1.0 is taken.

3. If the surveyed historical flood level in river reaches with steady but non-uniform flow, where there are obvious flood surface profile curves, the water surface profile may be calculated by the following equations to get historical flood discharge:

$$H_1 = H_2 + \frac{Q^2}{2}\left[\left(\frac{1}{K_1^2} + \frac{1}{K_2^2}\right)L - \frac{1\text{-}\xi}{g}\left(\frac{1}{A_1^2} - \frac{1}{A_2^2}\right)\right] \tag{6.3.1-9}$$

$$A_1 = A_{c1} + A_{t1} \tag{6.3.1-10}$$

$$A_2 = A_{c2} + A_{t2} \tag{6.3.1-11}$$

4. If the surveyed historical flood level is in river throat, where there is no scour in the riverbed, the historical flood discharge may be calculated by the following equation:

$$Q = A_2\sqrt{\frac{2g(H_1 - H_2)}{\left(1 - \frac{A_2^2}{A_1^2}\right) + \frac{2gLA_2^2}{K_1K_2}}} \tag{6.3.1-12}$$

where:

H_1, A_1 water level (m) and discharge area (m^2) at upstream cross-section of the river throat;
H_2, A_2 water level (m) and discharge area (m^2) at the cross-section of the river throat;
K_1, K_2 water delivery coefficient (m^3/s) at upstream cross section of the river throat and at detention spot section, respectively.

6.3.2 Empirical frequency of historical flood discharge may be calculated by the Eq. (6.2.3-2) in the *Specifications* after the order of the historical flood discharge is determined according to descriptions of long-term local residents or information in historical documents.

6.3.3 Design flood discharge may be calculated based on the following provisions:

1. Design flood discharge is calculated by the historical flood discharge which should not be less than twice. Values of C_V and C_S shall be consistent with regional distribution regularities. If the divergence is large, causes shall be analyzed and adjustments shall be made.
2. If there are many cases of historical flood discharge, which can be plot on Hessen probability grid paper to draw the empirical frequency curve, value of $\overline{Q}$, C_V, C_S and Q_P may be calculated according to Clauses 6.2.5 and 6.2.6 in the *Specifications.*
3. If the empirical frequency curve cannot be obtained from each historical flood discharge using the Hessen probability grid paper, the design flow may be calculated by the following methods:

(1) Selecting C_V, C_S with reference to the regional data.
(2) Calculating average flow using the following equations:

$$\overline{Q}_{Ti} = \frac{Q_{Ti}}{1 + \Phi_T C_V} \tag{6.3.3-1}$$

$$\overline{Q} = \frac{\sum_{i=1}^{n} \overline{Q}_{Ti}}{n} \tag{6.3.3-2}$$

where:

$\overline{Q}_{Ti}$ average discharge (m^3/s) calculated according to the *i*th historical flood discharge;
Q_{Ti} historical flood discharge (m^3/s) which is the *i*th flow whose recurrence period is *T* years;
Φ_T coefficient of deviation from mean with *T* years recurrence period;
n number of annual historical flood discharge cases.

(3) Calculating design flood discharge according to the Eq. (6.2.6) in the *Specifications.*

6.4 Other Methods for Calculating Design Flood Discharge

6.4.1 In regions without data, regional empirical equations and hydrological parameters may be used to calculate the design flood discharge, which shall be verified using the historical flood discharge.

6.4.2 For rivers where the catchment area is less than 100 km^2, the design flood discharge may be calculated by the rational equation, in which the parameters and indexes are obtained by graphs of storm runoff compiled in various regions.

6.4.3 Design flood discharge calculation for small bridges, subgrade works and special regions shall comply with the provisions in Chapters 9–11 of the *Specifications.*

6.5 Design Flood Level

6.5.1 If the river reach between the calculating section of bridge site and the hydrological cross-section is straight, the section is regular and the longitudinal gradient in river bottom is uniform, the relationship curve between water level and discharge at hydrological cross-section should be drawn according to the Eq. (6.3.1-1) in the *Specifications*; subsequently, the design flood level at the calculating section of bridge site is determined using the water surface gradient after the design flood level is determined by the design flood discharge at hydrological cross-section.

6.5.2 If there are blocks or artificial structures in upstream or downstream of the calculating section of bridge site and the hydrological cross-section, or there is significant difference between the shape and the area of the sections, and the longitudinal gradient of the riverbed has obvious curve, the water surface profile under the design flood discharge should be calculated using the trial-and-error method according to the Eq.(6.3.1-9) in the *Specifications* to calculate the design flood level.

6.5.3 Design flood level calculation for special regions shall comply with the provisions in Chapter 11 of the *Specifications.*

6.6 Design Flood Hydrograph

6.6.1 If observation data on flood discharge is available, measured flood hydrograph with greater flood which is adverse to bridge design may be selected as the example, then it is amplified to design flood hydrograph according to homogenous proportion. The amplification ratio may be calculated by the following equation:

$$k_{\mathrm{g}} = \frac{Q_{\mathrm{p}}}{Q} \tag{6.6.1}$$

where:

k_{g} amplification ratio;
Q peak discharge of typical flood (m^3/s);
Q_{p} design flood discharge (m^3/s).

6.6.2 If no flood observation data exist, the design flood hydrograph may be drawn by methods used by the local water conservancy department.

Chapter 7
Bridge Span Design

7.1 General

7.1.1 Bridge span design must ensure the safe passage of floods and silts of various classes no larger than the design flood, and meet the requirements of navigation, the passage of floating ice and other drifts.

7.1.2 Arrangement of bridge spans shall fit the characteristics and the change features of different types of river reaches, avoid unfavorable deformation of riverbed, and be economical and reasonable. Characteristics of different types of river reaches and riverbed evolution are given in Appendix A of the *Specifications.*

7.1.3 Backwater height in front of the bridge, changes of flow trend and deformation of riverbed after the construction of the bridge shall be within the allowable safety level.

7.1.4 Bridge spans shall be designed by considering the influences of riverbed evolution caused by the constructed or proposed hydraulic works, and navigation channel docks and pipelines in the upstream and downstream of the bridge site on the bridge spans.

7.1.5 The natural river channel near the bridge site should not be excavated or changed. If excavation and change of river channel is needed, reliable technical and economic evaluation shall be carried out.

7.1.6 Bridges that cross over estuaries, gulfs or link islands must meet the requirements of normal service and navigation under the influences of various marine hydrological conditions, such as tide, wave, storm surge, ocean current and sediment movement on seabed.

Y. Jiao, *Hydrological Specifications for Survey and Design of Highway Engineering*, JTG Standards and Specifications, https://doi.org/10.1007/978-981-92-3128-7_7

7.2 Bridge Span Length

7.2.1. The minimum length of clear spans should comply with the following provisions:

1. In the river reaches in canyons, bridge spans may be arranged according to the topography of the riverbed, and the channel should not be reduced by bridge. In this case, the minimum clear length of bridge span may not be calculated.
2. In open, straight or slightly bent, branching, and curved reaches or in unsteady reaches with separable floodplain and channel, the minimum length of clear spans should be calculated by the following equation:

$$L_j = K_q \left(\frac{Q_p}{Q_c} \right)^{n_3} B_c \qquad (7.2.1\text{-}1)$$

where:

L_j minimum length of clear spans (m);
Q_p design discharge (m^3/s);
Q_c channel discharge (m^3/s);
B_c channel width (m);
K_q, n_3 coefficient and index, which shall be adopted according to Table 7.2.1.

3. In a river reach with wide floodplain, the minimum clear length of bridge span should be calculated by the following equations:

$$L_j = \frac{Q_p}{\beta \cdot q_c} \qquad (7.2.1\text{-}2)$$

$$\beta = 1.19 \left(\frac{Q_c}{Q_t} \right)^{0.10} \qquad (7.2.1\text{-}3)$$

where:

β coefficient of flow compressibility;
q_c average discharge of unit width of channel [$m^3/(s \cdot m)$];
Q_t floodplain discharge (m^3/s).

Table 7.2.1 Values of K_q and n_3

Type of river reach	K_q	n_3
Open, straight or slightly bent reach	0.84	0.90
Branching or curved reach	0.95	0.87
Unsteady reaches with separable floodplain and channel	0.69	1.59

4. In an unsteady reach with floodplain and channel difficult to be divided, the minimum clear length of bridge span should be calculated by the following equations:

$$L_j = C_p \cdot B_0 \quad (7.2.1\text{-}4)$$

$$B_0 = 16.07\left(\frac{\overline{Q}^{0.24}}{\overline{d}^{0.3}}\right) \quad (7.2.1\text{-}5)$$

$$C_p = \left(\frac{Q_p}{Q_{2\%}}\right)^{0.33} \quad (7.2.1\text{-}6)$$

where:

B_0 basic channel width (m);
$\overline{Q}$ mean of annual maximum discharge (m^3/s);
$\overline{d}$ mean particle size of bed material (m);
C_p flood frequency coefficient;
$Q_{2\%}$ flood discharge with a frequency of 2% (m^3/s).

7.2.2 Design length of clear spans shall meet the requirements of the minimum clear length in Clause 7.2.1 of the *Specifications.* Besides, it shall be determined with comprehensive verification through technical and economic comparisons for different bridge lengths, according to the conditions such as topography of bridge site, riverbed geology, backwater in front of the bridge, scour depth, longitudinal gradients of the bridge and approach roads, and backfilling height, etc.

7.3 Arrangement of Bridge Spans

7.3.1 Arrangement of bridge span shall fit with the sectional discharge distribution of a natural river. In stable reaches, the ratio of the bridge spans lengths at the left and right floodplains shall be comparable to that of the discharges at the left and right floodplains; in sub-stable and unstable reaches, the influences of riverbed deformation and the variation trend of discharge distribution must be taken into account in the arrangement of bridge spans. The channel should not be reduced by the bridge spans, but the floodplains may be reduced appropriately.

7.3.2 In the navigable reach of inland river, arrangement of navigation spans shall meet the requirements of navigation clearance. And changes in the navigation channel caused by riverbed evolution and different water levels shall be fully taken into account.

7.3.3 In the river reach with floodwall, the disturbing of existing embankment shall be avoided for the arrangement of bridge spans. Emergency access for flood fighting should be provided in the intersection of the embankment.

7.3.4 Piers and abutments should not be set at the locations with poor geological conditions about fault, sink hole, karst cave and landslide, etc.

7.3.5 In the river reach with serious ice floe, the bridge span shall be enlarged appropriately, and the measures to prevent ice collision shall be provided.

7.3.6 Arrangement of bridge spans over mountainous rivers shall comply with the following provisions:

1. Single span should be set across the river reach in a canyon. Design elevation of deck shall be determined according to design flood level together with such conditions as topography and routes of both banks.
2. In the broad reaches, floodplains may be reduced by bridge spans appropriately. Embankments on floodplains should be perpendicular to the main flow direction of flood. Regulation works shall be supplemented if the bridge is skewed relative to the flow direction.
3. For a longitudinal bridge along a river in mountainous area, its route elevation should be raised, and it should be arranged at the foot of the mountain slope and to avoid water or occupy less water.

7.3.7 Arrangement of bridge spans for rivers in plain areas shall comply with the following provisions:

1. In the straight or slightly bent river reaches, the influences of downward movement of side floodplain and swinging of main channel within the river channel shall be taken into account for the arrangement of bridge spans.
2. In the curved river reaches, the development of river bend and the change of thalweg shall be predicted through investigation of riverbed evolution, and the influences of concentrated flow scour at the concave bank of river and the silting at the convex bank on the bridge spans and piers shall be taken into account.
3. In the branching reaches with stable floodplain and channel, if the discharge distribution is almost stable for many years, multiple bridges may be set over one river. Variation of discharge distribution ratio of each branch shall be estimated for arrangement of bridge spans, and the corresponding structures for river diversion shall be provided according to the discharge distribution ratios.
4. In the river reaches with wide floodplain, bridge spans may be arranged based on the mainstream trend and the swing range of thalweg at the upstream and downstream of the bridge site. The floodplains may be reduced the bridge spans appropriately, but the influences of backwater on the upstream shall be taken into account. If the river branch is stable but should not be channeled into the bridge spans, the construction of multiple bridges over one river may be considered.
5. In the meandering river reaches, the riverbed should not be reduced by bridge spans too much, and shall be designed in accordance with the local management and planning and with the support of regulation works.

7.3.8 Arrangement of bridge spans for the river in mountain-front areas shall comply with the following provisions:

1. In the river reach changed at the mountain-front, the floodplain with the supplemented suitable regulating structures may be reduced bridge spans significantly. Bridge axis shall be perpendicular to the bank line or the general trend of flood. Small bridges and culverts should not be constructed on the floodplain embankments. If the scheme of multiple bridges over one river is adopted, branches near the main channel shall be blocked.
2. In the alluvial and sheetflood reaches, bridge spans should cross over the narrow section in upstream or the contraction section in downstream of the river. In the reach with broad riverbed and obvious branches in the flow, the multiple bridges over one river scheme may be adopted, and the corresponding diverting and protective measures shall be provided between the bridges. The influence of the silting of riverbed shall be considered for the clearances beneath the bridges.

7.4 Design Elevation of Deck

7.4.1. Design elevation of deck for non-navigable rivers shall be calculated in accordance with the following provisions:

1. If the minimum deck elevation is calculated based on the design water level, the following equation shall be adopted:

$$H_{\text{min}} = H_{\text{S}} + \sum \Delta h + \Delta h_{\text{j}} + \Delta h_{\text{O}} \qquad (7.4.1\text{-}1)$$

where:

H_{min} minimum deck elevation (m);

H_{S} design water level (m);

$\sum \Delta h$ sum of various factors (m) such as backwater, wave height, wave set-up, superelevation of river bend, rising of local water surface in spate, rising of local water flow (only the larger value is considered between rising of local water surface in spate or local water flow), height of silting on riverbed, height of drift, etc.;

Δh_{j} safe value of clearance beneath bridge (m), which shall comply with the provisions in Table 7.4.1;

Δh_{O} construction height of superstructure of bridge (m), in which the height of deck pavement shall be included.

2. If the minimum deck elevation is calculated based on the design maximum floating ice water level, the following equation shall be adopted:

$$H_{\text{min}} = H_{\text{SB}} + \Delta h_{\text{j}} + \Delta h_{\text{O}} \qquad (7.4.1\text{-}2)$$

Table 7.4.1 Safe value of clearance beneath bridge for non-navigable rivers, Δh_j

Position in bridge	Safe value of clearance under bridge calculated by design water level (m)	Safe value of clearance under bridge calculated by maximum floating ice water level (m)
Bottom of girder	0.50	0.75
Top surface of bearing padstone	0.25	0.5
Arch springing	0.25	0.25

Note 1. Springing of hingeless arch may be submerged by flood, but the submerged depth should not be more than two-thirds of the rise of arch ring; clear height from the bottom of arch crown to the design water level shall not be smaller than 1 m

2. For the river in mountainous areas with significantly varying water level, the safe value of clearance beneath bridge may be increased appropriately

where:

H_{SB} design maximum floating ice water level (m), in which the height of silting on riverbed shall be considered.

3. Design elevation of deck shall not be smaller than the value calculated by the Eq. (7.4.1-1) or Eq. (7.4.1-2).

7.4.2 Design elevation of deck for navigable rivers shall meet the requirements for non-navigable rivers, and also comply with the following equation:

$$H_{\min} = H_{\text{tn}} + H_{\text{M}} + \Delta h_0 \quad (7.4.2)$$

where:

H_{tn} design maximum navigation water level (m);
H_{M} height of navigation clearance (m).

Chapter 8
Calculation for Scour at Piers and Abutments and Embedded Depth of Foundation

8.1 General

8.1.1 The scour of bridge piers and abutments shall include scour due to riverbed evolution, general scour and local scour. Calculation of pier and abutment scour shall be provided for the design basis to determine the embedded depth of foundation.

8.1.2 Calculation of pier and abutment scour shall be carried out using the corresponding methods and equations in the *Specifications* according to regional characteristics, features of river reach, hydrology and sediment characteristics, river geology, etc. If necessary, other equations or real measurement and survey data may be adopted for verification, and reasonable calculation results may be chosen after analysis and demonstration.

8.1.3 For special bridges where the hydrology and sediment conditions are complicated or the shape coefficient of pier is difficult to be determined, the scour depth may be determined by hydraulic model tests.

8.1.4 For bridges to be reconstructed or expanded, the adverse effects caused by their interference with the existing bridge in terms of flow direction, flow velocity and scour shall be taken into account for pier and abutment scour.

8.2 Scour Due to Natural Evolution of Riverbed

8.2.1 Based on investigations or data on riverbed section, topographic map of river reach, flood and sediment over years, the natural undercut of riverbed may be analyzed year by year, and the depth of the natural undercut during the service

Y. Jiao, *Hydrological Specifications for Survey and Design of Highway Engineering*, JTG Standards and Specifications, https://doi.org/10.1007/978-981-92-3128-7_8

life of the bridge may be estimated. This may also be estimated using a one-dimensional mathematical model of the riverbed silting as shown in Appendix B of the *Specifications*. The two approaches may then be compared and checked.

8.2.2 For the natural evolution scour caused by lateral changes in river channel, the calculated section should be selected as the section that is unfavorable in calculating the scour within the reach of bridge site.

8.2.3 The lowest elevation Z_w after maximum natural undercut in the riverbed of the concave bank in the river bend may be calculated by the following equations:

$$Z_w = Z_d - (1 + \xi)(Z_d - Z_b) \tag{8.2.3-1}$$

$$\xi = 2.07 - \lg\left(\frac{r_c}{B} - 2\right), \quad 2 < \frac{r_c}{B} < 22 \tag{8.2.3-2}$$

where:

Z_d design water level (m);
ξ shape coefficient of river bend;
Z_b average riverbed elevation formed under design discharge (m);
r_c radius of curvature at the top of river bend (m);
B natural river width (m).

8.2.4 The deformation of riverbed caused by the existing works associated with the river may be predicted by the existing analysis data and results of mobile-riverbed model test, or be determined by calculation with the corresponding equations

8.3 Calculation for General Scour Beneath Bridge

8.3.1 General scour of non-cohesive soil riverbed shall be separately calculated for channel and floodplain by the following equations

1. Channel

   (1) Simplified Eq. 64-2

$$h_p = 1.04\left(A_d \frac{Q_2}{Q_c}\right)^{0.90}\left[\frac{B_c}{(1-\lambda)\mu B_{cg}}\right]^{0.66} h_{cm} \tag{8.3.1-1}$$

$$Q_2 = \frac{Q_c}{Q_c + Q_{t1}} Q_p \tag{8.3.1-2}$$

$$A_d = \left(\frac{\sqrt{B_z}}{H_z}\right)^{0.15} \tag{8.3.1-3}$$

where:

h_p maximum water depth after general scour under the bridge (m).
Q_p design flood discharge (m^3/s).
Q_2 design flood discharge passing through the channel under the bridge (m^3/s). It is equal to Q_p if the channel can be widened to the whole bridge span.
Q_c design flood discharge of the channel in natural state (m^3/s).
Q_{t1} design flood discharge of the floodplain under the bridge in natural state (m^3/s).
B_c channel width in natural state (m).
B_{cg} channel width within the bridge length (m). It takes the total length of bridge spans if the channel can be widened to the whole bridge span.
B_z channel width corresponding to the dominant discharge (m). It may take the channel width at flat floodplain water level for compound riverbed.
λ the ratio between the total obstruction areas of piers to water flow and the discharge area within the width B_{cg} corresponding to the design water level.
μ lateral compression coefficient of flow around pier, which is determined according to Table 8.3.1-1.
h_{cm} maximum water depth of channel (m).
A_d discharge concentration coefficient of unit width. If $A_d > 1.8$ for mountain-front changing reach, meandering reach or wide floodplain reach, A_d may take 1.8.
H_z average water depth of channel corresponding to dominant discharge (m). It may take average water depth of channel at flat floodplain water level for compound riverbed.

(2) Modified Eq. 64-1

$$h_p = \left[\frac{A_d \frac{Q_2}{\mu B_{cj}} \left(\frac{h_{cm}}{h_{cq}} \right)^{\frac{5}{3}}}{E \bar{d}^{\frac{1}{6}}} \right]^{\frac{3}{5}} \tag{8.3.1-4}$$

where:

B_{cj} clear width of bridge span in channel (m). If the channel under the bridge can be widened to the whole bridge span, it takes the total clear width of all bridge spans.
h_{cq} average water depth in channel under the bridge (m).
$\bar{d}$ average particle diameter of sediment in channel (mm).
E coefficient related to sediment concentration in flood season, which may adopt the value in Table 8.3.1-2.

(3) General scour under the bridge may be estimated by numerical method based on the one-dimensional mathematical model of riverbed silting in Appendix B.

Table 8.3.1-1 Lateral compression coefficient of flow around pier, μ

Design velocity, v_s (m/s)	Clear length of single span, L_0 (m)							
	≤10	13	16	20	25	30	35	40	45
<1	1.00	1.00	1.00	1.00	1.00	1.00	1.00	1.00	1.00
1.0	0.96	0.97	0.98	0.99	0.99	0.99	0.99	0.99	0.99
1.5	0.96	0.96	0.97	0.97	0.98	0.98	0.98	0.99	0.99
2.0	0.93	0.94	0.95	0.97	0.97	0.98	0.98	0.98	0.98
2.5	0.90	0.93	0.94	0.96	0.96	0.97	0.97	0.98	0.98
3.0	0.89	0.91	0.93	0.95	0.96	0.96	0.97	0.97	0.98
3.5	0.87	0.90	0.92	0.94	0.95	0.96	0.96	0.97	0.97
≥4.0	0.85	0.88	0.91	0.93	0.94	0.95	0.96	0.96	0.97

Note 1. Coefficient μ refers to the reduction factor with flow area reduced due to stagnation zone caused by the vortex at the sides of piers and abutments

2. If clear length of single span $L_0 > 45$ m, $\mu = 1\text{–}0.375\,\frac{v_s}{L_0}$. For spans with unequal lengths, mean of μ values of all the spans may be adopted. If clear length of single span $L_0 > 200$ m, $\mu \approx 1.0$

Table 8.3.1-2 Coefficient relevant to sediment concentration in flood season, E

Sediment concentration, ρ (kg/m^3)	<1.0	1 ~ 10	>10
E	0.46	0.66	0.86

Note Sediment concentration ρ takes the average value of the maximum sediment concentration of the months in flood seasons over the years

2. Floodplain

$$h_p = \left[\frac{\frac{Q_1}{\mu B_{tj}}\left(\frac{h_{tm}}{h_{tq}}\right)^{\frac{5}{3}}}{v_{H1}}\right]^{\frac{5}{6}} \quad (8.3.1\text{-}5)$$

$$Q_1 = \frac{Q_{t1}}{Q_C + Q_{t1}} Q_p \quad (8.3.1\text{-}6)$$

where:

- Q_1 design discharge passing through the floodplain under the bridge (m^3/s);
- h_{tm} maximum water depth in the floodplain under the bridge (m);
- h_{tq} average water depth in the floodplain under the bridge (m);
- B_{tj} clear length of bridge span in the floodplain (m);
- v_{H1} flow velocity of non-cohesive soil without scour if the water depth in the floodplain is 1 m (m/s), which may adopt the value in Table 8.3.1-3.

Table 8.3.1-3 Flow velocity of non-cohesive soil without scour at water depth of 1 m

Sediment in riverbed		$\overline{d}$(mm)	v_{H1} (m/s)	Sediment in riverbed		$\overline{d}$(mm)	v_{H1} (m/s)
Sand	Fine	0.05–0.25	0.35–0.32	Cobble	Small	20–40	1.50–2.00
	Middle	0.25–0.50	0.32–0.40		Middle	40–60	2.00–2.30
	Coarse	0.50–2.00	0.40–0.60		Large	60–200	2.30–3.60
Pebble	Small	2.00–5.00	0.60–0.90	Boulder	Small	200–400	3.60–4.70
	Middle	5.00–10.00	0.90–1.20		Middle	400–800	4.70–6.00
	Large	10–20	1.20–1.50		Large	>800	>6.00

8.3.2 General scour of cohesive soil riverbed shall be calculated separately for channel and floodplain by the following equations

1. Channel

$$h_p = \left[\frac{A_d \frac{Q_2}{\mu B_{cj}} \left(\frac{h_{cm}}{h_{cq}}\right)^{\frac{5}{3}}}{0.33\left(\frac{1}{I_L}\right)}\right]^{\frac{5}{8}} \tag{8.3.2-1}$$

where:

A_d concentration coefficient of discharge of unit width, which takes 1.0–1.2;
I_L liquidity index of cohesive soil within the range of scour pit, which ranges in 0.16–1.19.

2. Floodplain

$$h_p = \left[\frac{A_d \frac{Q_1}{\mu B_{tj}} \left(\frac{h_{tm}}{h_{tq}}\right)^{\frac{5}{3}}}{0.33\left(\frac{1}{I_L}\right)}\right]^{\frac{6}{7}} \tag{8.3.2-2}$$

8.3.3 Approach velocity in front of the pier after general scour should be calculated by the following equations

1. When using Eq. (8.3.1-1) in the *Specifications* to calculate general scour:

$$v = \frac{A_d^{0.1}}{1.04}\left(\frac{Q_2}{Q_c}\right)^{0.1}\left[\frac{B_c}{\mu(1-\lambda)B_{cg}}\right]^{0.34}\left(\frac{h_{cm}}{h_c}\right)^{\frac{2}{3}} v_c \tag{8.3.3-1}$$

where:

v approach velocity in front of the pier after general scour (m/s);
v_c average flow velocity in channel (m/s);
h_c average water depth in channel (m).

2. When using Eq. (8.3.1-4) in the *Specifications* to calculate general scour:

$$v = \overline{Ed}^{\frac{1}{6}} h_{\mathrm{p}}^{\frac{2}{3}} \tag{8.3.3-2}$$

3. When using Eq. (8.3.1-5) in the *Specifications* to calculate general scour:

$$v = v_{\mathrm{H1}} h_{\mathrm{p}}^{\frac{3}{5}} \tag{8.3.3-3}$$

4. When using Eq. (8.3.2-1) in the *Specifications* to calculate general scour:

$$v = \frac{0.33}{I_{\mathrm{L}}} h_{\mathrm{p}}^{\frac{3}{5}} \tag{8.3.3-4}$$

5. When using Eq. (8.3.2-2) in the *Specifications* to calculate general scour:

$$v = \frac{0.33}{I_{\mathrm{L}}} h_{\mathrm{p}}^{\frac{1}{6}} \tag{8.3.3-5}$$

8.4 Calculation for Local Scour at Piers and Abutments

8.4.1 Local scour at piers and abutments in non-cohesive soil riverbed may be calculated by the following equations

1. Equation 65-2

   If $v \le v_0$

$$h_{\mathrm{b}} = K_{\xi} K_{\eta 2} B_1^{0.6} h_{\mathrm{p}}^{0.15} \left(\frac{v - v_0'}{v_0} \right) \tag{8.4.1-1}$$

   If $v > v_0$

$$h_{\mathrm{b}} = K_{\xi} K_{\eta 2} B_1^{0.6} h_{\mathrm{p}}^{0.15} \left(\frac{v - v_0'}{v_0} \right)^{n_2} \tag{8.4.1-2}$$

$$K_{\eta 2} = \frac{0.002\,3}{\overline{d}^{2.2}} + 0.375 \overline{d}^{0.24} \tag{8.4.1-3}$$

$$v_0 = 0.28 \left(\overline{d} + 0.7 \right)^{0.5} \tag{8.4.1-4}$$

$$v'_0 = 0.28 \left(\overline{d} + 0.5 \right)^{0.55} \tag{8.4.1-5}$$

$$n_2 = \left(\frac{v_0}{v}\right)^{0.23+0.19\lg\overline{d}} \quad (8.4.1\text{-}6)$$

where:

h_b local scour depth at pier and abutment (m);

K_ξ pier shape coefficient, which may take its value according to Appendix C in the *Specifications*;

B_1 alculated width of pier (m);

$\overline{d}$ average particle diameter of the sediment in riverbed (mm);

$K_{\eta 2}$ influence coefficient of particles in riverbed;

v approach velocity in front of the pier after general scour (m/s), which may be calculated according to Clause 8.3.3 in the *Specifications*;

v_0 critical velocity of the sediments on the riverbed (m/s);

v_0' approach velocity if particles at a pier begin to move (m/s);

n_2 index.

2. Modified Eq. 65-1

If $v \le v_0$

$$h_b = K_\xi K_{\eta 2} B_1^{0.6}(v - v_0') \quad (8.4.1\text{-}7)$$

If $v > v_0$

$$h_b = K_\xi K_{\eta 1} B_1^{0.6}(v - v_0')\left(\frac{v - v_0'}{v_0 - v_0'}\right)^{n_1} \quad (8.4.1\text{-}8)$$

$$v_0 = 0.0246\left(\frac{h_p}{\overline{d}}\right)^{0.14}\sqrt{332\overline{d} + \frac{10 + h_p}{\overline{d}^{0.72}}} \quad (8.4.1\text{-}9)$$

$$K_{\eta 1} = 0.8\left(\frac{1}{\overline{d}^{0.45}} + \frac{1}{\overline{d}^{0.15}}\right) \quad (8.4.1\text{-}10)$$

$$v_0' = 0.462\left(\frac{\overline{d}}{B_1}\right)^{0.06} v_0 \quad (8.4.1\text{-}11)$$

$$n_1 = \left(\frac{v_0}{v}\right)^{0.25\overline{d}^{0.19}} \quad (8.4.1\text{-}12)$$

where:

$K_{\eta 1}$ influence coefficient of particles in riverbed;

n_1 index;

$\overline{d}$ average particle diameter of bed material, which ranges in 0.1–500 mm;

h_p the maximum water depth under bridge after general scour, which ranges in 0.2–30 m;

v approach velocity in front of pier after general scour, which ranges in 0.1–6 m/s;

B_1 calculated width of pier, which ranges in 0–11 m.

8.4.2 Local scour at piers and abutments in cohesive soil riverbed may be calculated by the following equations

If $\frac{h_p}{B_1} \geq 2.5$

$$h_b = 0.83K_\xi B_1^{0.6} I_L^{1.25} v \tag{8.4.2-1}$$

If $\frac{h_p}{B_1} < 2.5$

$$h_b = 0.55K_\xi B_1^{0.6} h_p^{0.1} I_L^{1.0} v \tag{8.4.2-2}$$

where:

I_L liquidity index of cohesive soil within the range of the scour pit, which ranges in 0.16–1.48.

8.4.3 Maximum scour depth at abutments shall be determined after analysis, calculation and comparison according to riverbed characteristics, contraction degree of the bridge site, etc. Local scour depth at piers and abutments in non-cohesive soil riverbed may be calculated separately for channel and floodplain by the following equations

1. For an abutment located in channel:

If $\frac{h_p}{\bar{d}} \leq 500$

$$h_b = 1.17k_\varepsilon k_\alpha h_p \left(\frac{l}{h_p}\right)^{0.6} \left(\frac{\bar{d}}{h_p}\right)^{-0.15} \left[\frac{(v - v_0')^2}{g h_p}\right]^{0.15} \tag{8.4.3-1}$$

If $\frac{h_p}{\bar{d}} > 500$

$$h_b = 1.17k_\varepsilon k_\alpha h_p \left(\frac{l}{h_p}\right)^{0.6} \left(\frac{\bar{d}}{h_p}\right)^{-0.10} \left[\frac{(v - v_0')^2}{g h_p}\right]^{0.15} \tag{8.4.3-2}$$

$$k_\alpha = \left(\frac{\alpha}{90}\right)^{0.2}, \quad \alpha \leq 90° \tag{8.4.3-3}$$

where:

h_b local scour depth at abutments (m).

k_ε abutment shape coefficient, which may adopt the value in Table 8.4.3.

α angle between bridge (abutment) axis and flow direction. If the bridge axis is perpendicular to the flow direction, $\alpha = 90°$.

Table 8.4.3 Abutment shape coefficient, k_ε

Shape of abutment	k_ε	Shape of abutment	k_ε
Embedded vertical abutments	0.39–0.42	Spill-through abutments	0.43–0.47
Gravity U-shape abutments	0.92		

k_α coefficient of the angle between abutment axis and flow direction. The application range of is 0°–90°, which is calculated by Eq. (8.4.3-3).

l length of abutment and embankment which are perpendicular to flow direction, or the width of abutment and embankment obstructing the flow (m). Its application range is $\frac{l}{h_p} = 0.16 - 8.80$.

h_p water depth at floodplain under bridge after general scour (m).

$\overline{d}$ average particle diameter of floodplain sediment (m).

v approach velocity in front of the abutment after general scour (m/s), which may be calculated according to Clause 8.3.3 in the *Specifications*.

v_0' initial scour velocity of sediment in front of the abutment (m/s), which may be calculated according to the Eq. (8.4.1-9) and Eq. (8.4.1-11) in the *Specifications*.

g 9.80(m/s).

2. If the abutments are located in floodplain, local scour depth may be calculated by Eqs. (8.4.3-1) and (8.4.3-3), in which the variables of water and sediment take the corresponding values at floodplain.

8.5 Scour Calculation in Special Situations

8.5.1 For the layered riverbed under the bridge that is composed of several layers of soil with different materials, scour of the riverbed may be calculated by the layer-by-layer method.

8.5.2 Scour of rock may be calculated on the basis of the classification of rock specified in Appendix D of the *Specifications*.

8.6 Minimum Embedded Depth of Pier and Abutment Foundations

8.6.1 The embedded depth of bridge pier and abutment foundations shall be determined using the unfavorable combination of scour due to natural evolution of riverbed, general scour and local scour, which shall be selected according to the

Table 8.6.2 Safe value for embedded depth of foundation (m)

Total scour depth (m)		0	5	10	15	20
Safe value	Large, medium, small bridges (without pavement)	1.5	2.0	2.5	3.0	3.5
	Super large bridges	2.0	2.5	3.0	3.5	4.0

Note 1. Total scour depth is the sum of scour due to natural evolution of riverbed, general scour and local scour calculated from the riverbed surface

2. The values in the table are the minimum embedded depth of pier and abutment foundations below the total scour depth. The value may be increased properly without accurate data of design discharge, water level and original section or if accurate data about riverbed evolution cannot be acquired
3. If there are existing bridges or reconstructed bridges in the upstream or downstream of the bridge site, scour by catastrophic flood of existing bridge shall be investigated, and safe value shall be added to maximum scour depth of the existing bridge as the embedded depth of pier and abutment foundations of the new bridge

specific conditions of the river reach at the bridge site. Also, the determination shall comply with the current relevant provisions in *Ground Base of Highway and Bridge and Basic Design Specifications* (JTG D63).

8.6.2 Safe value for the embedded depth of pier and abutment foundation in non-rock riverbed may be determined according to Table 8.6.2.

8.6.3 The minimum embedded depth of pier and abutment foundations in rock riverbed shall be determined according to Appendix D in the *Specifications* by considering the possible scour on rock, and in terms of rock hardness, binding category, weathering degree, fissuring and its development.

8.6.4 For the abutment located in the channel, if its maximum scour depth is smaller than the total scour depth of piers, the embedded depth of abutment foundation shall be the same as the elevation of the pier foundation. For the abutment located in floodplain, if the river is unstable with swinging channel, the elevation of abutment foundation shall be the same as the pier. For stable rivers, the elevation of abutment foundation may be determined based on the calculation result of abutment scour.

8.6.5 In determining the embedded depth of the footing of the abutment cone slope, the influence of scour shall be considered. If the abutment is located in the floodplain of stable or sub-stable river reach, the bottom of the foundation shall be at least 0.5 m below the general scour level. If the abutment is located in the floodplain of unstable river reach, the bottom of the foundation shall be at least 1 m below the general scour level.

Chapter 9
Hydrological Survey and Design of Small Bridges and Culverts

9.1 Principles of Arrangement

9.1.1 In accordance with terrain, geology, hydrology and other conditions along the route and considering the drainage system in the whole route, small bridges and culverts shall be laid out economically and properly to adapt to irrigation and drainage of farmland, and to achieve the drainage ability corresponding to the design flood frequency, which is specified in Clause 1.0.8 of the *Specifications*.

9.1.2 The location of small bridges and culverts shall meet the demand of alignment along the route. If it is not restrained by the alignment, the small bridges and culverts should be located in the river (ditch) reaches where there are favorable terrain, good geological conditions, foundation with high bearing capacity and stable riverbed.

9.1.3 Small bridges and culverts shall be set on every catchment area or drainage river channel. If the local terrain condition allows and the technology and economy are reasonable, one bridge may be set to accommodate a group of channels.

9.1.4 If a small bridge and a culvert are close to the downstream import watercourse, the influence of the design water level and the changes of scouring and silting of the downstream watercourse on the clear height and foundation embedded depth of the bridge and culvert shall be considered.

9.1.5 In the alluvial fan area of the mountain pass, small bridges and culverts shall be set separately, and should not be set at the low-lying place. Small bridges and culverts shall be set on the low-lying land between two alluvial fans.

9.1.6 In the region of sheet flood without apparent channels, the flood should be discharged by sections. Small bridges and culverts shall be set over different streams, but should not be overcrowded together.

9.1.7 Small bridges and culverts in irrigation regions shall be arranged in cooperation with the drainage-irrigation systems in farmlands. If a minor change of the existing

Y. Jiao, *Hydrological Specifications for Survey and Design of Highway Engineering*, JTG Standards and Specifications, https://doi.org/10.1007/978-981-92-3128-7_9

drainage-irrigation systems is necessary, the original drainage-irrigation function shall not be degraded.

9.1.8 Spans of small bridges and culverts over the drainage-irrigation ditch may be determined according to the design discharge cross-section of the drainage-irrigation ditch. Spans of small bridges and culverts over natural river ditches may be drafted according to the cross-sectional shapes of the river ditches and be checked in accordance with the provisions in Sect. 9.4 of the *Specifications*. The natural drainage areas of the river ditches under the criteria of design flood and drainage areas of river channel should not be reduced by drafted spans excessively.

9.1.9 The influence of salivary flow ice shall be considered in the design of span and height for small bridges and culverts in cold areas.

9.1.10 The arrangement of entrance and exit shall facilitate the drainage of water. If necessary, diversion or drainage works may be provided to match with the entrance and exit.

9.1.11 For the submersible small bridges and culverts or fords under a flood frequency of 1/25 on Class-3 highways, the safe movement of vehicles shall be guaranteed, and water depth on the bridge deck or roadway for vehicle traffic shall not be larger than 0.3 m. For submersible small bridges and culverts or fords on Class-4 highways, the traffic may be interrupted to some extent under a flood frequency of 1/25, and the time period of interruption may be determined on specific situations.

9.2 Hydrological Investigation and Survey

9.2.1 The following information shall be collected before hydrological investigation and survey

1. Topographic map along a river.
2. Data required for calculation of design discharge, including mean annual precipitation over several years, capacity and intensity of 24 h rainfall corresponding to design flood frequency, etc.
3. Calculation methods for regional flood, records of historic flood, and existing flood calculation results in each stream.
4. Existing drainage-irrigation systems and drawings of planning scheme, design cross-section, discharge and water level of each drainage-irrigation ditch, etc.

9.2.2 Hydrological investigation and survey shall include the following contents

1. Soil classification, vegetation conditions, distribution and current conditions of water storage work in each catchment area.
2. Roughness of the riverbed shall be determined according to the soil classification at both banks of the river channel and the riverbed material.

3. If a bridge (culvert) is located near a village, historic flood levels, normal water levels, scouring and silting on the riverbed and drifters shall be investigated.
4. Current situation, structural type, embedded depth of foundation, variation of scour and operation of the existing bridges and culverts shall be investigated. The situation of salivary flow ice shall also be investigated for cold regions in north.
5. Survey of gradient of river channel shall be conducted. The range for survey shall be subject to the calculation for the gradient of the river channel where the bridge (culvert) is located. In plain terrain, the range is not smaller than 200 m at the upstream of the hydrological cross-section and 100 m at the downstream of the hydrological cross-section. In mountain areas, the range is not smaller than 100 m at the upstream and 50 m at the downstream.
6. Arrangement and survey of hydrological cross-section shall be conducted. If the route is skewed with the river channel, the hydrological cross-section shall be arranged and surveyed near the bridge (culvert). If the historic flood level is distant from the bridge (culvert) and there is great variation of the cross-section of the river channel, the hydrological cross-section shall also be arranged and surveyed near the historic flood level. The range of survey is subject to the water level and the calculated discharge.

9.3 Hydrological Calculation

9.3.1 Design discharge of small river basins in mountainous and hilly areas may be calculated according to the provisions in Clause 6.4.2 of the *Specifications* or the empirical equations for the regional discharge. Various methods shall be used for comparison and verification, and reasonable calculation results shall be adopted by comprehensive analysis.

9.3.2 Design discharge of small river basins in plain terrain should be calculated by using the empirical equations for the regional discharge or according to the method in Section 6.3 of the *Specifications*. If only one historical flood level can be surveyed, its return period shall be determined in accordance with the conditions of the regional historical flood.

9.3.3 Within the same hydrological subregion, if there are similar catchment areas or reliable results of design discharge in the same catchment area, or there is flood information to reliably calculate the design discharge, the design discharge at the bridge (culvert) location may be calculated according to the Eq. (6.2.2) in the *Specifications*.

9.3.4 For river channels with available historical flood levels through survey, the design discharge calculated according to various equations shall be verified by the historical flood discharge.

9.3.5 The design water level corresponding to the design discharge may be obtained by a trial-and-error method or plotting of the relationship between water level and discharge according to the Eq. (6.3.1-1) in the *Specifications.*

9.4 Span Design

9.4.1 Spans of small bridges and culverts must be designed to ensure safe passage of design flood and drifters, satisfy the need of irrigation and drainage, avoid the adverse effects on farmlands and buildings in the upstream and downstream areas, as well as take into account the economic justification of the project cost.

9.4.2 Small bridges should be designed to be of a non-free out-flow state, and culverts shall be designed to be of inlet unsubmerged type. The safe value of the clearance under a bridge shall be in accordance with the provisions in Table 7.4.1. The clearance from the top of an inlet unsubmerged culvert to the maximum flow surface shall be in accordance with the provisions in Table 9.4.2. The water depth in front of a culvert shall be no larger than 1.15 times the clear height of the culvert.

9.4.3 In span calculation for small bridges and culverts, the effect of water logging in front of the bridges and culverts on the design discharge may be ignored.

9.4.4 Checking calculation for the span of a small bridge over the river with paved bed may be carried out by drafting the span according to the cross-section shape of the river channel, and then the flow velocity, the water depth as well as the backwater height may be checked.

9.4.5 Calculating the span of a small bridge may be carried out by drafting the span according to the cross-section shape of the river channel, and then calculating the backwater height, checking the backwater level in front of the bridge and the clearance under the bridge, calculating the scour depth, as well as checking the embedded depth of the foundation according to the provisions in Chapter 7 and 8 of the *Specifications.* The safe value of the embedded depth of the foundation shall not be less than 1 m.

9.4.6 Checking the calculation for the span of an inlet unsubmerged culvert may be carried out by drafting the span according to the cross-section shape of the river

Table 9.4.2 Clear height of inlet unsubmerged culvert

Clear height at entrance of culvert, h_d (m)	Culvert type		
	Pipe culvert	Arch culvert	Rectangular culvert
≤ 3	$\geq h_d/4$	$\geq h_d/4$	$\geq h_d/6$
>3	≥ 0.75 m	≥ 0.75 m	≥ 0.50 m

channel, and then checking the flow velocity, the water depth under the bridge as well as the backwater height in front of the culvert.

9.5 Protection Against Scour

9.5.1 Protection measures against scour should be taken in the upstream and downstream of the small bridges and culverts and within a certain range of the subgrade slope.

9.5.2 If the longitudinal gradient of the channel floor is smaller than or equal to 15%, the longitudinal gradient of the paving of the bridge and culvert may be similar to that of the channel floor; while the longitudinal gradient of the channel floor is larger than 15%, the paving of the bridge and culvert should be made into a step type according to the channel slope or set as a longitudinal slope with its gradient no larger than the critical one, and also smoothly connected with the natural river channel.

9.5.3 If the longitudinal gradient of the channel floor is smaller than or equal to 15%, the paving range of protection on the river floor of the bridge and culvert should be extended to the ends of wing walls at the upstream and downstream, and shall also be provided with cut-off walls at the ends of the paving surfaces at the upstream and downstream. The embedded depth of the cut-off wall shall not be less than that of the abutment or the wing walls. If the out-flow velocity of the bridge and culvert is larger than the flow velocity allowable by the soil of the riverbed, a flip bucket shall be provided on the paving surface at the downstream outlet of the culvert, in which the type of the flip bucket may be determined by the length of the paving, or stone rubbles are dump-filled at the end of the downstream paving surface. The elevation of the paving surface should be slightly lower than that of the riverbed, and the type of the paving shall match with the design flow velocity.

Chapter 10
Hydrological Survey and Design of Subgrade

10.1 General

10.1.1 Hydrological survey and design of subgrade shall be carried out by applying corresponding survey and design methods based on regional characteristics and hydrological features, and providing hydrological basis for reasonable design of surface drainage and protection works of the subgrade.

10.1.2 The design flood frequency for the surface drainage and protection works of the subgrade shall comply with the provisions in Clause 1.0.8 of the *Specifications*.

10.1.3 Hydrological survey and design of subgrade should be carried out in line with that of the bridge and culvert, and the design of the drainage system along the route is also taken into account comprehensively.

10.2 Hydrological Investigation and Survey

10.2.1 In normal areas, general situations of catchment areas shall be investigated, rainstorm runoff and geological data shall be collected, and scopes of salivary flow ice and surface water together with exposing location, discharge, seasonal change of groundwater shall be investigated.

10.2.2 Hydrological investigation and survey for subgrade along the river shall comply with the following provisions:

1. Contents of hydrological investigation should be planned in accordance with the provisions in Section 5.2 of the *Specifications*.
2. Spacing between hydrological cross-sections arranged along the route may be 1–2 times of the river width. Hydrological cross-sections shall be added in the

Y. Jiao, *Hydrological Specifications for Survey and Design of Highway Engineering*, JTG Standards and Specifications, https://doi.org/10.1007/978-981-92-3128-7_10

locations where there are obvious changes in the cross-section and the longitudinal gradient of the riverbed or feeding of large branches, in the front and back of the bypass flow, and near the survey points for historic flood levels. Measurement range of the hydrological cross-sections shall comply with the provisions in Clause 5.3.1 of the *Specifications*.
3. Topographic map of river reaches shall include the reaches that affect the project, and the elevation shall be surveyed above the historic maximum flood level. The contents of the survey focus on the topography and ground objects that influences the flow direction, and the hydrological cross-section and historical flood level shall be marked.
4. The slope of riverbed may be plotted by using the topographic map of river reaches. Surveying and mapping of the flood slope shall comply with the provisions in Clause 5.3.2 of the *Specifications*.

10.2.3 Hydrological investigation and survey for subgrade in the floodplain may be carried out in accordance with the provisions in Chapter 5 of the *Specifications*.

10.2.4 In the low-lying (river network) areas of plains, hydrological investigation and survey for subgrade may be carried out by complying with the provisions in Section 11.3 of the *Specifications*.

10.2.5 Hydrological investigation and survey for subgrade in coastal regions may be carried out by complying with the provision in Clause 11.7.1 of the *Specifications*.

10.3 Hydrological Analysis and Calculation

10.3.1 In normal areas, design discharge may be calculated according to the provisions in Section 6.4 of the *Specifications*, while the depth of surface run-off or water-logging may be determined by survey.

10.3.2 Hydrological analysis and calculation for subgrade along the river shall be carried out according to following provisions:

1. Design discharge may be derived according to the provisions in Chapter 6 of the *Specifications*.
2. Design water surface profile may be determined by the design water level derived according to the relation curve of water level versus discharge of the hydrological cross-section, together with the riverbed slope and the historical flood slope; the water surface profile method may also be used for the design water surface profile.
3. The average velocity of the channel or the floodplain near the subgrade may be used at the water flow velocity along the subgrade, but the effects caused by the possible riverbed evolution shall be taken into account. If the subgrade is impacted by the skew flow, the effects of non-uniform distribution shall be taken into account in the flow velocity along the subgrade and may be calculated by the following equation:

$$v_\beta = \frac{2\varepsilon_\beta}{1+\varepsilon_\beta} v_c(t) \tag{10.3.2}$$

where:

v_β velocity of skew flow (m/s);
$v_c(t)$ average velocity of the channel or the floodplain near the subgrade (m/s);
ε_β coefficient of non-uniform distribution of flow velocity, which may take values in Table 10.3.2.

4. The subgrade along the river should not be intruded into the river channel, and the discharge area of the riverway should not be reduced by it. The angle between the subgrade and the flow should not be too large. If the reduction of the discharge area is too large, the height of backwater shall be calculated to analyze the influences on the channel direction, riverbed deformation, embankment and farmhouse.

10.3.3 Hydrological analysis and calculation for the subgrade on the floodplain shall be calculated according to the following provisions

1. Design discharge, design water level and flow velocity on the floodplain shall be obtained from the hydrological analysis and calculation for the bridge site.
2. The maximum backwater height in the upstream side of the subgrade without diversion dike or with pear-shaped diversion dike at the bridge ends may be determined by following provisions:

(1) The maximum backwater height may be calculated by the following equations:

$$\Delta h_{sh} = \Delta Z + L_{y1} I \tag{10.3.3-1}$$

$$L_{y1} = K_s(1 - M')B \tag{10.3.3-2}$$

where:

Δh_{sh} maximum backwater height above the design water level in upstream side of the subgrade (m);
ΔZ maximum backwater height in front of a bridge (m);
L_{y1} distance between the location of the maximum backwater height in front of a bridge and the bridge axis (m);
I natural flood slope of river reach in the bridge site (in decimal term);
M' ratio between the discharge passing through the bridge span range in the natural state and the design discharge;

Table 10.3.2 Coefficient of non-uniform distribution of flow velocity, ε_β

Angle between flow and subgrade, β	≤15°	20°	30°	40°	50°	60°	70°	80°	90°
ε_β	1.00	1.25	1.50	1.75	2.00	2.25	2.50	2.75	3.00

B water surface width under the design flood (m);

K_s coefficient, which may take the value in Table 10.3.3-1.

(2) The distance between the position of the maximum backwater height and the front edge of the abutment may be calculated by the following equations:

$$L_{sh} = AL_{y1} - 0.5L_d \qquad (10.3.3\text{-}3)$$

$$E' = 1 - \frac{Q'_{t2}}{Q'_{t1}} \qquad (10.3.3\text{-}4)$$

where:

L_{sh} distance between the position of the maximum backwater height and the front edge of the abutment (m);

L_d distance between the front edges of two abutments (m);

A coefficient, which may take the value in Table 10.3.3-2 according to E' and M';

E' bridge span offset factor;

Q'_{t1} lager discharge obstructed by the subgrade in one end of the bridge (m^3/s);

Q'_{t2} smaller discharge obstructed by the subgrade in the other end of the bridge (m^3/s).

(3) Transverse gradient of water surface along the subgrade, and transverse gradient of water surface from the front edge of the abutment to the position of the maximum backwater height may be calculated by the following equation. The water surface between the position of the maximum backwater height and the bank is calculated by the flat slope.

$$I_h = \frac{L_{y1}I}{L_{sh}} \qquad (10.3.3\text{-}5)$$

Table 10.3.3-1 Coefficient, K_s

M'	0.8	0.7	0.6	0.5
K_s	0.45	0.49	0.53	0.59

Table 10.3.3-2 Coefficient, A

M'	E'									
	0	0.1	0.2	0.3	0.4	0.5	0.6	0.7	0.8	0.9
0.5	1.43	1.44	1.48	1.55	1.63	1.73	1.85	1.98	2.14	2.31
0.6	1.93	1.94	1.95	2.00	2.09	2.20	2.35	2.52	2.73	2.97
0.7	2.80	2.81	2.82	2.83	2.90	3.03	3.23	3.47	3.79	4.16
0.8	4.60	4.64	4.68	4.72	4.77	4.87	5.16	5.57	6.16	6.92

(4) If the calculated L_{sh} is larger than the length of the subgrade on the floodplain, the maximum backwater height may be calculated by the following equation:

$$\Delta h'_{sh} = \Delta Z + \frac{L_a L_{y1} I}{L_{sh}} \tag{10.3.3-6}$$

where:

$\Delta h'_{sh}$ maximum backwater height (m) above the design water level at the junction of the side slope and the bank slope in upstream of the subgrade, if $L_{sh} > L_a$;

L_a subgrade length (m) from the front edge of the abutment to the bank of the same end.

3. The backwater height in the upstream side of the subgrade where a non-enclosed diversion dike is set at the bridge ends should be determined by the following provisions:

(1) If $L_{sh} \le L_a$, the backwater height is calculated by Δh_{sh};
(2) If $L_{sh} > L_a$, the backwater height is calculated by $\Delta h'_{sh}$;
(3) The water surface profile along the subgrade is regarded as a flat slope.

4. The water levels on the downstream side of the subgrade is calculated using the same value. The decrease between this water level and the design water level may be estimated by the following equation:

$$\Delta h_x = K_j h_{t1} \tag{10.3.3-7}$$

where:

Δh_x decrease of water level (m);
h_{t1} average water depth within the range of the subgrade in the floodplain at the design water level (m);
K_j factor for water level decrease, which may take the value in Table 10.3.3-3.

5. The maximum difference of water levels at the upstream and downstream sides may be calculated by the following equations:

$$\Delta h_m = \Delta h_{sh} + \Delta h_x \tag{10.3.3-8}$$

or

$$\Delta h_m = \Delta h'_{sh} + \Delta h_x \tag{10.3.3-9}$$

where:

Δh_m maximum difference of water levels at the upstream and downstream sides (m).

Table 10.3.3-3 Factor for water level decrease, K_j

$\frac{Q_t'}{Q_p}$	E'											
	One end of the subgrade in floodplain with larger obstructing discharge						One end of the subgrade in floodplain with smaller obstructing discharge					
	0	0.2	0.4	0.6	0.8	1.0	0	0.2	0.4	0.6	0.8	1.0
0	0.00	0.00	0.00	0.00	0.00	0.00	0.00	0.00	0.00	0.00	0.00	0.00
0.1	0.07	0.08	0.10	0.12	0.13	0.14	0.07	0.07	0.07	0.06	0.06	0.06
0.2	0.13	0.17	0.20	0.23	0.25	0.26	0.13	0.13	0.12	0.12	0.11	0.11
0.3	0.19	0.25	0.29	0.33	0.35	0.36	0.19	0.19	0.18	0.18	0.18	0.17
0.4	0.25	0.33	0.38	0.41	0.43	0.44	0.25	0.24	0.24	0.23	0.23	0.22
0.5	0.30	0.40	0.44	0.46	0.48	0.48	0.30	0.29	0.28	0.27	0.26	0.24
0.6	0.33	0.42	0.47	0.49	0.51	0.51	0.33	0.32	0.30	0.29	0.28	0.26
0.7	0.36	0.44	0.49	0.51	0.52	0.52	0.36	0.34	0.32	0.30	0.28	0.27

Note 1. Value of K_j listed in the table may be calculated by interpolation
2. Q_t' in the table is the total discharge obstructed by two ends of the subgrade (m^3/S), E' is the bridge span offset factor, Q_p is design discharge (m^3/s)

6. Flow velocities along the upstream and downstream sides of the subgrade may be determined by the equation as follows. The flow velocity under the bridge should be adopted as the flow velocities near abutments.

$$v_j = 0.7 v_{tj} \quad (10.3.3\text{-}10)$$

where:

v_j flow velocity along the upstream and downstream sides of the subgrade (m/s);

v_{tj} average flow velocity within the range of the subgrade in the floodplain in natural state (m/s).

10.3.4 In the plain low-lying (river network) areas, hydrological analysis and calculation shall be conducted according to the following provisions

1. For the scenarios with flood diversion or detention in the upstream of a subgrade:
   (1) Design discharge and water level may be calculated by the principle of water balance according to the discharge hydrograph at the design flood frequency in the inlet of flood diversion and the catchment areas in the upstream side of the subgrade, the relationship line of water storage volume versus water level in the upstream of the subgrade, and the relationship curve of water level and discharge at the hydrological cross-section of the subgrade.
   (2) If control valves are provided for flood discharge in outlets, the influences on the discharge at the subgrade cross-section shall be taken into account in the water balance calculation.

2. For the scenarios without flood diversion or detention in the upstream of a subgrade:

   (1) The design discharge in the subgrade section may be calculated by the following equation:

$$Q_{p2} = Q_p - Q_{pl} \tag{10.3.4}$$

   where:

   Q_{p2} design discharge in the subgrade section (m^3/s);
   Q_p total design discharge in the catchment areas (m^3/s), which may be calculated by rainfall data or regional empirical equation;
   Q_{pl} design discharge for channel flood drainage in catchment areas (m^3/s).

   (2) Design water level may be estimated by the Eq. (11.3.2-1) in the *Specifications* or the survey data according to the characteristics of low-lying areas.

10.3.5 Hydrological analysis and calculation for the coastal subgrade may be carried out by complying with the provisions in Clause 11.7.2 and 11.7.3 of the *Specifications*

10.4 Height of Submerged Subgrade

10.4.1 The height of submerged subgrade shall comply with the provision of the minimum height of earth filling specified in the *Design Standard for Highway Subgrade* (JTG D30), and the design elevation of its edges shall be higher than the value calculated by the following equation

$$H_{min} = H_S + \sum \Delta h + 0.5 \tag{10.4.1}$$

where:

H_{min} minimum elevation for the subgrade edge (m);
H_S design flood level (m);
$\sum \Delta h$ sum of backwater height or decrease value of water level, wave run-up, local rising-up, super elevation of river bend and silting-up of riverbed, etc.

10.4.2 The backwater height of the subgrade along the river or in the floodplain shall be calculated according to the Clause 10.3.2 and 10.3.3 of the *Specifications.* Wave run-up and silting-up of riverbed may be determined by calculation or survey. If the subgrade along the river is located in the concave bank or the flow direction is skew with the axis of the subgrade, local rising-up should be calculated by the following equation, and the calculated value is compared with the wave run-up value and the larger one is taken for the Eq. (10.4.1) in the *Specifications.*

$$\Delta h_{\mathrm{ch}} = \frac{v_{\mathrm{g}}^2 \sin^2\theta}{g\sqrt{1+m^2}} \tag{10.4.2}$$

where:

Δh_{ch} local rising-up of skew flow on the slope of the subgrade (m);
v_{g} average velocity of the flow rushing to the subgrade (m/s);
θ skew angle between the flow direction and the tangent line of river bend or the axis of the subgrade axis (°);
m slope factor of the subgrade surface facing water.

10.4.3 For the subgrade in the floodplain with enclosed diversion dike, if the design criteria of the diversion dike is lower than the design flood frequency, the minimum elevation of the subgrade edges shall be calculated by the Eq. (10.4.1) of the *Specifications*; if the diversion dike is strong enough to withstand the flood of the design frequency, the decrease of water level may be calculated by the Eq. (10.3.3-7) and replay the backwater height in the Eq. (10.4.1) of the *Specifications*, in which the wave run-up and the silting-up of riverbed are determined by calculation or survey.

10.4.4 The backwater height, wave run-up and silting-up of riverbed in the subgrade section in plain low-lying (river net) areas shall be determined by calculation or survey.

10.4.5 The design elevation of the coastal subgrade shall not be lower than the highest tide level of the design flood frequency plus the height of wave attack and the safety height of 0.5 m.

10.5 Protection Against Scour

10.5.1 If the submerged subgrade is scoured by water flow, suitable protection measures and necessary regulation works shall be selected according to the river feature, the property of water flow, the landform of river channel, the geology, etc., together with the location of the subgrade.

10.5.2 For the coastal subgrade, suitable protection measures and necessary regulation works shall be selected according to the characteristic of the wave action to the subgrade, the water depth in front of embankment, the wave height, the maximum bottom flow velocity of wave, and the conditions of sub soil, etc.

10.5.3 For the direct protection works subjected to the scour from the water flow, the footing shall be embedded no less than 1.0 m below the scouring line or embedded into bed rock. Surface protection measures may also be used.

10.5.4 The scour depth of the direct protection works for the subgrade along the river shall include the following three parts:

1. Scour due to natural evolution of riverbed, which should be determined according to Section 8.2 of the *Specifications.*
2. General scour, which should be calculated according to Section 8.3 of the *Specifications.*
3. Local scour, which should be determined through calculation together with the field investigation and analysis.

Chapter 11
Hydrological Survey and Design of Bridges in Special Areas

11.1 Reservoir Areas

11.1.1 Data described as follows shall be collected if bridge site is in the area influenced by reservoir impoundment:

1. Nature, grade, design standard of the reservoir and the dimension of main dam, auxiliary dam and spillway.
2. Topographic maps, profile and cross section maps of the reservoir.
3. Relationship curves of water level and volume, water level and water surface area, water level and total outflow of the reservoir.
4. Relationship curves of water level and discharge, water level and velocity, and hydrograph of design flood discharge and historic maximum discharge in natural state.
5. Average discharge and water volume, and grading curve of suspending sediments, which flow into the reservoir in flood season over the years, etc.
6. All kinds of characteristic water level, backwater curve, design deposit years and deposit calculating data of reservoir, etc.
7. Maximum wave height, wave intrusion height, wave length and corresponding data about water level of reservoir.
8. Ice condition of the reservoir area.
9. Possible highest and lowest water level of the reservoir area and duration during the bridge construction.
10. Reconstruction and expansion planning of reservoir.

11.1.2 The calculation of design flood discharge should comply with the following provisions if bridge site is in the area influenced by reservoir impoundment:

1. Design flood discharge may be calculated according to natural state if the bridge is located in fluctuating backwater section of the reservoir.

Y. Jiao, *Hydrological Specifications for Survey and Design of Highway Engineering*, JTG Standards and Specifications, https://doi.org/10.1007/978-981-92-3128-7_11

2. If the design flood frequency of the reservoir exceeds that of bridge criterion, inflow hydrograph that is in agreement with design flood frequency of the bridge shall be adopted to calculate pondage and design flood discharge. Initial adjusting water level shall be determined by the service condition of the reservoir.
3. If the checked flood frequency of reservoir is below that of bridge criterion, the maximum discharge running through the bridge in reservoir at dam failure shall be checked according to the equations given below:

$$Q_{\mathrm{k}} = Q_{\mathrm{kh}} \frac{W_{\mathrm{q}}}{W - W_{\mathrm{h}}} \tag{11.1.2-1}$$

$$Q_{\mathrm{kh}} = 0.27\sqrt{g}\left(\frac{L_{\mathrm{k}}}{B_{\mathrm{k}}}\right)^{\frac{1}{10}}\left(\frac{B_{\mathrm{k}}}{b_{\mathrm{k}}}\right)^{\frac{1}{3}} b_{\mathrm{k}}\left(h_{\mathrm{b}} - k_{\mathrm{O}}h'_{\mathrm{b}}\right)^{\frac{3}{2}} \tag{11.1.2-2}$$

$$L_{\mathrm{k}} = \frac{W \cdot 10^4}{h_{\mathrm{b}} B_{\mathrm{k}}} \tag{11.1.2-3}$$

If $W > 10^6\ \mathrm{m}^3$

$$b_{\mathrm{k}} = K_1 K_2 W^{\frac{1}{4}} B_{\mathrm{k}}^{\frac{1}{7}} h_{\mathrm{b}}^{\frac{1}{2}} \tag{11.1.2-4}$$

If $W \le 10^6\ \mathrm{m}^3$

$$b_{\mathrm{k}} = K_1 K_3 (W h_{\mathrm{b}})^{\frac{1}{4}} \tag{11.1.2-5}$$

$$k_{\mathrm{O}} = 1.4\left(\frac{b_{\mathrm{k}} h'_{\mathrm{b}}}{B_{\mathrm{k}} h_{\mathrm{b}}}\right)^{\frac{1}{3}} \tag{11.1.2-6}$$

where:

Q_{k} maximum discharge running through the bridge in the reservoir after dam-failure occurs (m^3/s).
Q_{kh} maximum discharge running through the dam after dam-failure occurs (m^3/s).
W_{q} capacity in front of the bridge if dam failure occurs ($10^4\ \mathrm{m}^3$).
W total capacity in upstream side of the dam if dam failure occurs ($10^4\ \mathrm{m}^3$).
W_{h} capacity of residual height of the dam ($10^4\ \mathrm{m}^3$).
L_{k} length of reservoir (m), the length from dam section to upstream end of the reservoir where submerged width narrows abruptly may be adopted, or it may be calculated by Eq. (11.1.2-3); if $\frac{L_{\mathrm{k}}}{B_{\mathrm{k}}} > 5$, $\frac{L_{\mathrm{k}}}{B_{\mathrm{k}}} = 5.0$.
B_{k} length of dam (m).
b_{k} average opening of failed dam (m); it is equal to the dam length for concrete gravity dam; it may be calculated by Eqs. (11.1.2-4) and (11.1.2-5) for common earth embankment dam and rock-fill dam; it is equal to the dam length if the calculated value is larger than the dam length.

h_b water depth in upstream side of the dam if dam-failure occurs (m), which may take the value of dam height if check-calculating normal reservoir.

h'_b residual height of the dam after failure (m), which may be disregarded if estimation is difficult.

k_0 modification coefficient, which takes 0.94 if the calculated value is greater than 0.94.

K_1 safety coefficient, which takes 1.1–1.3 according to highway class and conditions of dam quality.

K_2 building material coefficient of the dam, which takes 1.2 for clay, clay core wall, sloping core or dam made of earth, rock and concrete, and 2.0 for homogeneous loam.

K_3 material quality coefficient, which takes 6.6 for fine quality, and 9.1 for poor quality.

11.1.3 Design flood level shall comply with following provisions if the bridge site is in an area influenced by reservoir impoundment:

1. Design flood level may be determined by using the maximum impounded level under the bridge corresponding to the design flood frequency of the bridge offered by the water conservancy department; or by the method of 'computation of reservoir backwater'. The influence of silting-up of the reservoir shall be taken into account in the design flood level.
2. Design flood level shall be determined using the maximum impounded level under the bridge corresponding to the checked flood frequency of reservoir if the checked flood frequency of the reservoir is smaller than bridge design flood criterion.

11.1.4 Computation of reservoir backwater shall comply with the provisions given below if the bridge site is in an area influenced by reservoir impoundment:

1. Reservoir backwater curve may be derived using the principle of energy balance or water balance and based on reservoir profile, cross section and inflow discharge hydrograph that is consistent with the design flood frequency of the bridge.
2. Profile and cross section of the reservoir shall take the sections with reservoir sedimentation during the service period of the bridge; while the calculation section is selected at the changed place of the reservoir topography.
3. Inflow discharge hydrograph which is in line with design flood frequency of the bridge may be calculated according to data of the reservoir or to Chapter 6 in the *Specifications.*
4. If determining the water level in front of dam and the outflow by calculation of flood regulation, the initial regulating level shall be determined according to operation of the reservoir during flood season, and normal impounded level of reservoir or limiting level during flood season may be adopted.
5. Backwater curve corresponding to that discharge at each time interval shall be calculated according to the inflow discharge hydrograph of the design flood. After comparing the water level under the bridge of each backwater curve, the maximum value shall be taken as the design flood level.

11.1.5 Data given below shall be collected if bridge site is at downstream of a reservoir:

1. Basic data such as construction time, design standard, control catchment area, silting-up conditions of the reservoir, and outflow at the same design flood frequency of the bridge.
2. Maximum depth of local scour of river bottom at downstream of dam-apron and range of scour extending toward downstream.
3. Design flood discharge and catchment area of various tributaries between the bridge and the dam, in addition to the distance from the bridge to the dam, and the catchment area in-between.
4. Planning of water conservancy facilities near the bridge site.

11.1.6 Calculation of design flood discharge shall comply with provisions given below if the bridge site is at downstream of the reservoir:

1. If the bridge is close to the dam and their in-between catchment area is less than 10% of the catchment area upstream of the dam site, the outflow of the reservoir at the same flood frequency may be adopted as the design flood discharge under the bridge.
2. If the catchment area between the bridge and the dam is less than 20% of the catchment area upstream of the dam site and no more than 1000 km^2, the design flood discharge under the bridge may be estimated by the equation as below:

$$Q_{\mathrm{p}} = Q_{\mathrm{px}} + Q'_{\mathrm{p}}\left(\frac{F_{\mathrm{q}}}{F}\right)^{n_1} \tag{11.1.6-1}$$

where:

Q_{p} design flood discharge under the bridge (m^3/s);
Q_{px} outflow of the reservoir if flood of design frequency occurs (m^3/s);
Q'_{p} discharge of bridge-location section under the design flood frequency in natural state (m^3/s);
F catchment area upstream of the bridge (km^2);
F_{q} catchment area between the bridge and the dam (km^2);
n_1 area index.

3. If the catchment area between the bridge and the dam is no less than 20% of the catchment area upstream of the dam site or more than 1, 000 km^2, the design flood discharge under the bridge may be calculated by the equations as below, and the larger result value shall be taken:

$$Q_{\mathrm{p}} = Q'_{\mathrm{p}} - Q_{\mathrm{pb}} + Q_{\mathrm{px}} \tag{11.1.6-2}$$

$$Q_{\mathrm{p}} = Q_{\mathrm{pq}} + \left(Q'_{\mathrm{p}} - Q_{\mathrm{pq}}\right)\frac{Q_{\mathrm{px}}}{Q_{\mathrm{pb}}} \tag{11.1.6-3}$$

where:

Q_{pb} discharge of dam-location section under the same frequency of the bridge in natural state (m^3/s);

Q_{pq} discharge of catchment area between the bridge and the dam under design flood frequency (m^3/s).

4. The outflow of the reservoir may be acquired from the water conservancy department or determined by calculation of flood regulation. As for low-capacity reservoirs with flood regulation function that are short of data, and their design criteria conform to the required design criteria for highway bridges, the outflow may be calculated approximately by the equation as below:

$$Q_{px} = Q_{pb}\left(1 - \frac{W_F}{W_{bs}}\right) \tag{11.1.6-4}$$

where:

W_{bs} flood volume corresponding to the discharge of dam-location section (m^3);

W_F flood control volume of the reservoir (m^3).

5. If the design flood criterion of the bridge is higher than the checking criterion of the reservoir, the design flood discharge under the bridge shall not only be calculated according to the natural state, but also shall be checked by the following equation after the occurrence of dam failure:

$$Q'_k = \frac{W_{ks}}{\frac{W_{ks}}{Q_{kh}} + \frac{L_{bq}}{\bar{v}K_0}} \tag{11.1.6-5}$$

where:

Q'_k maximum discharge of bridge-location section after dam failure occurs (m^3/s).

W_{ks} dischraged water volume after dam failure occurs (m^3).

L_{bq} distance between the bridge and the dam (m).

$\bar{v}$ average velocity of maximum section of the riverway during flood season (m/s), maximum measured value may be adopted. As for areas without data: 3.0–5.0 m/s may be adopted for mountainous areas, 2.0–3.0 m/s for mountain front areas and 1.0–2.0 m/s for plain areas.

K_0 adjustment coefficient, 1.0–1.5 for mountainous areas, 1.0 for mountain front areas, 0.8–0.9 for plain areas.

11.1.7 If the bridge site is in downstream of the reservoir, calculation of design flood level may be determined according to the provisions in Section 6.5 of the *Specifications*. Original section before the reservoir construction should be used as the calculating section. If the riverbed section after the reservoir construction is adopted, design flood level should be raised according to scour conditions of the riverbed after the reservoir construction.

11.1.8 Design of bridge spans shall comply with following provisions:

1. If the bridge site is in the area influenced by reservoir impoundment:
   (1) In the reservoir fluctuating backwater zone, design length of bridge spans may be dealt with according to the natural state, while elevation of bridge deck may be determined by referring to the provisions in Section 7.4 of the *Specifications*; as for ice-frozen river, effects of level raise caused by ice dam and ice jam shall be taken into consideration.
   (2) Within the reservoir area, design length of bridge spans may be determined according to topographic conditions and economic comparison, and the elevation of deck may be determined according to the provisions in Section 7.4 of the *Specifications.*
2. If the bridge site is in downstream of the reservoir, bridge spans may be designed in accordance with provisions in Chapter 7 of the *Specifications.*

11.1.9 Calculation of scour at piers and abutments and embedded depth shall comply with the following provisions:

1. If the bridge site is in an area influenced by reservoir impoundment:
   (1) In the reservoir fluctuating backwater zone, embedded depth may be dealt with according to natural state.
   (2) Within the reservoir area, scour under the bridge may be calculated using equations in Chapter 8 of the *Specifications* with consideration of the influences caused by encounter of lowest impounded level of the reservoir during flood season and design flood, source-tracing scour of the reservoir, deformation of river caused by changes of water and sand condition after the construction of the reservoir, and scour caused by water discharge of the upstream reservoir.
2. If the bridge site is at downstream of the reservoir, the embedded depth of the abutment foundation shall comply with relevant provisions in Chapter 8 of the *Specifications,* in addition, the scour effects caused by discharged clear water of the reservoir shall be taken into account. Bridge site shall be selected beyond local scour zone at downstream of the dam, otherwise the scour depth shall be analyzed and calculated according to operation conditions of the reservoir and empirical equations, or determined by hydraulic physical modeling if necessary.

11.2 Debris Flow Areas

11.2.1 Hydrological investigation shall include the following contents:

1. Collecting the data such as regional rainstorm (intensity and duration), impoundment of the river basin (cover area and thickness of ice and snow, pondage of mountainous lakes, impoundment of reservoir), annual runoff modules, modules of flood-peak flow and data of regional topography, geology, soil, vegetation.
2. Investigating distribution and scales of unfavorable geological potential hazards of collapse, landslide, etc. within catchment areas, and estimating the amount of possible breakouts.
3. Investigating frequency, time, scale, process of debris flow within the catchment area, and nature of debris flow, flood mark, height of the first part of debris flow, riverbed gradient, flushing variations of the riverbed in catchment area.
4. Investigating morphological feature of the debris flow, historic vicissitudes of villages nearby, human activities and detriments of debris flow.

11.2.2 Hydrological calculation of debris flow reach shall comply with the following provisions:

1. Measuring unit weight of debris flow, if the representative samples of debris flow during flowing can be collected, the weighing method may be adopted; otherwise, the weight may be determined by calculating the ratio between solid matter and water of the debris flow by means of field investigation. Other empirical survey methods based on the investigation information may also be adopted.
2. The velocity of debris flow may be calculated using the empirical equation derived with measured or investigated data, or other equations that shall be compared with regional calculation equations for the debris flow velocity. If there are signification differences in the calculation results, rationality shall be analyzed.
3. Discharge of debris flow may be calculated by rainfall-flood correction method or by method of mud crack survey if encountering ditches with mud cracks. If necessary, both methods mentioned above may be used for calculation and cross checking.

11.2.3 In the deposit areas of debris flow course, silting-up of the riverbed shall be taken into account for design opening height of bridges and culverts. Expected total silting-up thickness accumulation during design years may be determined according to the quantity of loose material in the source of debris flow. Also, investigation of silting slope, length, erosion base, drainage base of stabilized debris flow may be carried out for comprehensive analysis to estimate the variation of the possible ultimate silting-up thickness. Thus, the ultimate silting-up thickness of the channel for debris flow, through which the route passes, may be deduced as the design total silting-up thickness, which may also be calculated with the following equations:

$$H_{\mathrm{n}} = K_2 N_{\mathrm{S}} \overline{h}_{\mathrm{n}} \tag{11.2.3-1}$$

$$\overline{h}_{\mathrm{n}} = \frac{\sum h}{N} \tag{11.2.3-2}$$

where:

H_n total silting-up thickness within design life (m);
K_2 silting-up trend coefficient, 1.0–1.5 is adopted during development stage of debris flow, 0.5–1.0 during decline stage;
N_S design period (a);
$\overline{h}_n$ average silting-up thickness of years (m/a);
$\sum h$ total silting-up thickness in N years of survey period, calculated by data of survey, excavation or drilling;
N survey period (a).

11.2.4 Bridge opening over the stream of debris flow shall be designed comprehensively by taking into account the section where the bridge is located, topographic conditions, channel width, properties of debris flow, streaming trend, development and changes of debris flow, and shall also comply with the following provisions:

1. For bridges cross over the passing through area of debris flow, the gully bed shall not be reduced by the bridge and piers should not be located in the gully.
2. For bridges cross over the transition area between the circulation zone and the deposit zone of debris flow, part of the gully may be reduced by the bridge spans. With the reduced gully bed, regulation works shall be constructed. By choosing river reach in upstream area with uniform channel width and steady flow as the basically steady reach, the minimum clear length of bridge spans may be calculated by the following equations:

$$L_j = \frac{n_2 I_1^{\frac{1}{2}}}{\mu n_1 I_2^{\frac{1}{2}}} B_1 \tag{11.2.4-1}$$

$$\mu = 1 - 0.375 \frac{\gamma_C \overline{v}}{10 L_0 + (\gamma_C - 10)} \tag{11.2.4-2}$$

where:

L_j minimum clear length of bridge spans (m);
B_1 ditch width of basically steady reach in upstream area (m);
n_1, n_2 coefficient of roughness of the basically steady reach in upstream area and bridge site reach, respectively;
I_1, I_2 slope of basically steady reach in upstream area and bridge reach (in decimal term), respectively;
μ lateral reduction factor of flow around piers and abutments;
γ_C unit weight of debris flow (kN/m^3);
$\overline{v}$ average flow velocity of debris flow section (m/s);
L_0 length of single span of the bridge (m).

3. Bridge opening should cover main flow ranges of debris flow if the bridge is over a deposit zone. As for dilute debris flow, whose flow and velocity is not great in margin areas of overflow groove, bridge spans may be reduced appropriately but should no more than 10%–20%, and drainage works shall be conducted.

11.2.5 Elevation of the bridge deck and road shoulder shall be calculated according to the following provisions:

1. Elevation of the bridge deck shall be calculated by the following equation:

$$H_{min} = H_b + h_{mc} + H_n + \Delta h_n + \Delta h_O + \Delta h_j \quad (11.2.5\text{-}1)$$

where:

H_{min} lowest elevation of the deck (m).
H_b average elevation of the riverbed under design mudflow level (m).
h_{mc} mean flow depth of debris flow corresponding to the design flood discharge (m), its value shall be no less than 1.5 times the wave height of sinuous flow and 1.5 times the diameter of maximum rock block.
H_n design total silting-up height (m), 25-year total silting-up height is adopted for design flood frequency of 1/25; 35-year total silting-up height is adopted for design flood frequency of 1/50; and 50-year total silting-up height is adopted for design flood frequency of 1/100.
Δh_n value of super elevation on curved ramps (m).
Δh_O structure depth of the bridge superstructure (including thickness of deck pavement) (m).
Δh_j safe value (m), 1.0 m may be adopted, or larger value for areas where intensive debris flow occurs or as conditions permit.

2. Lowest elevation of shoulder of the approach road shall be calculated by the following equation:

$$H_{min} = H_n + H_S + 0.5 \quad (11.2.5\text{-}2)$$

where:

H_{min} lowest elevation of road shoulder (m);
H_S elevation of mud flow surface of design flood frequency (m).

11.2.6 Scour at abutments may be determined according to field survey and analysis. For dilute debris flow, it may be calculated according to the following provisions:

1. At the section with straight groove, it may be calculated by the following equation:

$$h_p = \frac{0.1 q_s}{d^{\frac{1}{3}} h_m^{\frac{1}{6}}} \quad (11.2.6\text{-}1)$$

where:

h_p water depth after general scour (m);
q_s maximum unit width discharge of debris flow [$m^3/(s\ m)$];
d average grain size of sediments on the channel riverbed (m);
h_m maximum mud depth (m).

2. At the section with concave bank in curve course, it may be calculated by the following equation:

$$h_p = \frac{0.17q_s}{d^{\frac{1}{3}} h_m^{\frac{1}{6}}} \quad (11.2.6\text{-}2)$$

11.3 Low-Lying (River Network) Areas in Plain

11.3.1 Hydrological investigation shall include following contents:

1. Researching the history, current situation and management planning of the river basin. Collecting the planning reports about river courses, embankments, pump stations, culvert gates, flood control, flood diversion, flood drainage of the basin, and design data of profile, cross section of the channel and large-scale topographic map.
2. Investigating relevant inter-basin planning, distribution of water network and polder districts. Investigating the connection between polder districts, river forks and main water course, flow regulating direction; flood detention (storage) area, flood diversion area and distribution, design standards and operation principles of the dike locks; conditions of embankment and dyke failure, flood storage, flood diversion in history, various characteristic level of water.
3. Investigating silting conditions of heavily silt-carrying rivers in bottomland, and know about the capability of desilting and water-diversion.
4. As for a river equipped with flood control dam, embankment standard near the bridge site, and its maximum safety discharge flow, flood control level, cross dimensions and construction quality shall be investigated; investigating the highest historical flood level and maximum historical discharge through embankments; investigating conditions of embankment failure, dyke failure and overflow; collecting data such as drainage flow through the embankment outlet and drainage of pumping station; measuring the highest, lowest elevation of the levee crest and highest impounding elevation of the reach within embankment area; knowing about planning for improving embankment standard, etc.
5. Within waterlogging areas, investigation shall include topographic geomorphology and geological condition, farmland irrigation system in polder districts, flood storage and detention conditions of channels and lakes, water conservancy facility, drainage capacity of the pump lock, jacking process on drainage channels by outer rivers and outer seas; highest waterlogging level and its duration and precipitation in corresponding years.
6. In irrigation areas, investigation contents shall mainly include distribution of irrigation system, directions of regulated flow, means of drainage and irrigation, water resources, etc. As for large main channel and branch channel, section dimensions, maximum irrigation flow and corresponding data of water level shall be collected from water conservancy department. As for channels with

flood discharge functions, maximum flood discharge and corresponding water level shall be collected.

7. As for river network and navigation channel with comprehensive utilizations, grade of navigation channel, highest and lowest navigation level, highest design flood level and corresponding data about velocity and discharge shall be collected; navigation planning and desilting conditions of the navigation channel shall be investigated.
8. Collecting the data such as flood detention elevation, chart showing relationship between area and capacity, rate of inflow and outflow, outflow curve of the water diversion lock and various measured hydrological data in flood storage and detention area. If there is planning of temporary flood diversion by blasting embankment in upstream flood storage and detention area of the bridge, data such as its location, conditions of flood diversion, diversion volume, peak-down level after diversion, waterlogging elevation of flood storage and detention area if bombing, breach section and elevation of the lowest point shall be collected.

11.3.2 Hydrological calculation shall comply with following provisions:

1. Design flood level of the waterlogging areas may be calculated by the following equations, and the design flood discharge of waterlogging may be determined by local flood drainage equations or by data from the water conservancy department:

$$H_S = H + \Delta H \tag{11.3.2-1}$$

$$\Delta H = \frac{F\Delta h}{A_j} \tag{11.3.2-2}$$

where:

- H_S design flood level (m);
- H historic highest waterlogging level (m), determined by field survey;
- ΔH differential value between design flood level and historic highest waterlogging level (m);
- Δh difference between precipitation under design frequency and precipitation corresponding to historic highest waterlogging level (m), annual maximum precipitation under design frequency is compared with annual maximum precipitation corresponding to historic maximum waterlogging level, and the largest differential value is adopted;
- F catchment area of the basin (m^2);
- A_j historic maximum waterlogging area (m^2), which is determined by surveying waterlogging limits in large scale topographic map.

2. Design flood level and discharge of diversion, discharge and navigation channel may comply with the following provisions:

(1) Maximum diversion discharge or maximum flood drainage flow and corresponding flood levels supplied by the department managing the cannel may

be adopted as the design flood discharge and flood level of the diversion and drainage channel. They may be calculated using the equation for open-channel uniform flow if there is no data related to discharge and water level.

(2) In navigation channel, if there is man-made construction for regulating water level, maximum discharge and corresponding water level supplied by the department in charge may be adopted as the design flood discharge and level; if there is no man-made construction for regulating water level, the design flood discharge may be calculated by the equation for open-channel uniform flow and with the highest water level investigated.

3. Design flood level and discharge in flood detention (storage) areas, or in flood diversion course, should be calculated according to the provision in Clause 10.3.4 of the *Specifications*.
4. The maximum safety flood discharge flow and the corresponding water level shall be adopted for design flood discharge and level in river course with embankment. If there is a plan to upgrade the design criteria, the plan may be taken into account.

11.3.3 Design of bridge spans shall be determined by the following methods according to geographical environment and various hydrological conditions at the bridge site:

1. As for bridges spanning over flood drainage channels in waterlogging areas, bridge spans shall be arranged according to design flood discharge and level. Discharge areas should not be reduced by bridge spans and the need for long-term development shall be taken into account. Design elevation of deck shall not be lower than the calculated value specified in the provisions in Section 7.4 of the *Specifications*.
2. Discharge areas should not be reduced by bridge spans over diversion channels or flood discharge channels. Elevation of beam soffit shall be higher than the elevation of channel top which is to be raised. Requirements of navigation opening shall be taken into account for navigation channels.
3. Length of bridge spans over flood detention areas shall be calculated by combining the water balance equation and the bridge-span outflow equation; length of bridge spans in flood basin shall be calculated by considering that the discharge after waterlogging in front of the bridge is reduced; length of bridge spans over diversion channel shall be the sum of current basin discharge and flood diversion discharge. Discharge areas of main streams and flood drainage channels shall not be reduced by bridge spans. Backwater height in front of bridge should not exceed the allowable value. Design elevation of the deck shall not be lower than calculated value in accordance with the provisions in Section 7.4 of the *Specifications* and elevation of beam bottom shall not be lower than the crest elevation.
4. Minimum clear length of bridge spans, which are over the channel with flood control embankment, shall be controlled by allowable backwater height in front of the bridge. Some reserved length shall be taken into account to the bridge under

which the channel may be broadened in the long run. Elevation of beam soffit shall be higher than crest elevation in long-term planning. Navigation requirements shall be taken into account for channels which are open to navigation.

5. If affected by man-made structures or backwater jacking caused by downstream rivers and lakes, bridge spans shall be checked according to backwater jacking conditions and attention shall be paid to the effects of sand silting-up.

11.3.4 Scour depth under the bridge may be determined according to the provisions and equations in Chapter 8 of the *Specifications*. Requirements of removing salt and alkali shall be taken into account if determining the embedded depth of foundations, and for those channels which may be turned into flood drainage channels in the long run, effects of decreased riverbed elevation caused by flood control and exploiting water course shall also be considered. For bridges over channels for diversion, drainage, flood drainage or irrigation, the crest level of foundation should be 0.20 m lower than the elevation of the channel bottom. Minimum embedded depth of abutment floor may be determined according to the provisions in Section 8.6 of the *Specifications*.

11.4 Karst Regions

11.4.1 Hydrological investigation shall include following contents:

1. In catchment area, investigating surface water system, catchment area, underground watershed and its relationship with surface watershed.
2. Investigating location, shape, scale and intercepted catchment area of various karsts form such as inner dissolved spring, aven, dissolved basin, karst lake and underground river in catchment area. Finding out conditions of inflow and outflow of underground karst water and its effects on floods in bridge reach.
3. Measuring the flood discharge hydrograph, relation curve of level and flow and waterlogging level change hydrograph in the upstream and downstream of karst region.
4. Investigating the historic flood conditions in bridge reach and precipitation in the catchment area over the years.
5. Investigating the damage, deformation, renovation, remediation and operation of existing structures in karst areas along the route.

11.4.2 Hydrological calculation shall comply with the following provisions:

1. Discharge and discharge hydrograph at the design flood frequency, which are not affected by karst, may be determined according to the provisions in Chapter 6 of the *Specifications*. Flood volume may be determined after the enlargement of measured typical discharge hydrograph or by the following equation:

$$W'_{\mathrm{p}} = 0.1\alpha H_{24\mathrm{p}} F \tag{11.4.2-1}$$

where:

W_p' water volume under design flood frequency (10^4 m^3);

α runoff coefficient, data compiled by water conservancy department may be used;

H_{24p} maximum precipitation under design frequency in 24 hours (mm), data compiled by water conservancy department may be used;

F catchment area (km^2).

2. Surface and subsurface runoff distribution coefficient may be determined by the following equation:

(1) Surface runoff distribution coefficient:

$$p = \frac{Q_b}{Q_l} \quad (11.4.2\text{-}2)$$

or

$$p = \frac{W_b}{W_l} \quad (11.4.2\text{-}3)$$

where:

p surface runoff distribution coefficient;

Q_l peak discharge before water absorption of various karst (m^3/s);

Q_b surface peak discharge after water absorption of various karst (m^3/s);

W_l flood volume before water absorption of various karst (m^3);

W_b surface flood volume after water absorption of various karst (m^3).

(2) Ground water runoff distribution coefficient:

$$p' = 1 - p \quad (11.4.2\text{-}4)$$

where:

p' ground water runoff distribution coefficient.

3. Water absorption capability of karst may be calculated by the following equation:

$$\overline{Q}_x = \overline{Q}_l - \overline{Q}_b - \frac{\Delta W_j}{\Delta t} \quad (11.4.2\text{-}5)$$

where:

$\overline{Q}_x$ water absorption capability of various karst in any time interval Δt (m^3/s);

$\overline{Q}_l$ average inflow volume before water absorption of various karst (m^3/s) in time interval Δt, which may be obtained using the measured inflow discharge hydrograph in upstream side;

$\overline{Q}_{b}$ average surface downstream discharge after water absorption of various karst (m^3/s) in time interval Δt, which may be obtained using the measured surface discharge hydrograph in downstream side;

ΔW_j increased waterlogging volume in time interval Δt (m^3), which may be acquired by referring to relationship curve of waterlogging level H_j and waterlogging volume W_j;

Δt calculation time interval (s).

4. Determining design flood discharge and level according to the following provisions:

(1) If there are various karsts in catchment area and water yield in closure-area can be completely diverted underground under the design flood, the design flood discharge may be calculated by adopting catchment area in which closure-area is deducted.

(2) If there are various karst in catchment area and water yield in closure-area can not be completely diverted underground under design flood, the design flood discharge may be calculated by the following equation:

$$Q_p = Q'_p - Q_{plP'} \tag{11.4.2-6}$$

where:

Q_p design flood discharge (m^3/s);

Q'_p design flood discharge free from karst influence (m^3/s);

Q_{pl} inflow water yield under design flood frequency in upstream of karst (m^3/s).

(3) If there is water from the outlet of karst cave, spring and underground water which comes from other catchment areas, water yield may be measured by making rectangular or triangular weir at water outlet. Water yield may be added to the flow of catchment area directly to get the design flood discharge.

(4) If there are karst forms with functions of reducing flood crest such as karat ditches and cracks rather than obvious karst caves in catchment area, the design flood discharge may be reduced accordingly.

(5) Design flood level may be determined by the provisions in Section 6.5 of the *Specifications*. If the route passes through waterlogging low-lying lands such as polje and karst lake, design waterlogging level may be inferred after waterlogging volume is calculated by the following equation. The calculated level inferred shall be compared with surveyed historic flood level, and it shall be analyzed and checked if the difference is too large.

$$W_j = \sum \Delta t(\overline{Q}_{li} - \overline{Q}_{bi} - \overline{Q}_{xi}) \tag{11.4.2-7}$$

where:

W_j waterlogging volume (m^3);

$\overline{Q}_{\mathrm{l}i}$ average inflow discharge during time interval Δt on design flood discharge hydrograph (m^3/s);

$\overline{Q}_{\mathrm{b}i}$ downstream surface discharge acquired from relation curve of measured waterlogging level H_j and downstream surface discharge Q_b (m^3/s);

$\overline{Q}_{\mathrm{x}i}$ absorbed water volume acquired from relation curve of waterlogging level H_j and water absorption capability Q_x (m^3/s).

11.4.3 Arrangement of bridge spans shall be determined by the distribution and development of caves, grooves, hopper, and subsurface rivers. The location of piers should avoid karst spot, and shall be bypassed in areas with severe karst. Length of bridge spans and the bridge deck design elevation may be determined according to the provisions in Sections 7.2 and 7.4 of the *Specifications.*

11.4.4 Calculations of scour at piers and abutments and embedded depth of foundation may be determined according to the provisions in Chapter 8 of the *Specifications.*

11.5 Reach with Backward Flow

11.5.1 The following contents shall be included for hydrological investigation:

1. Collecting the data over the years about water level, discharge, velocity in upstream of bridge site where there is no backward flow effect, the highest backward flood level, riverbed slope, rate of rise and fall of backward flow flood, flushing variation, etc. in bridge reach.
2. Collecting topographic map of backward flow area and draw relationship curve between impounding water level in front of the bridge and water surface area.

11.5.2 The following three flow conditions shall be considered for hydrological calculation:

1. If large rivers have no backward flow effect on the bridge site, flood discharge and level under the bridge corresponding to design flood frequency may be calculated according in Chapter 6 of the *Specifications.*
2. If the river where the bridge is located has its regular level, while flood at design frequency occurs in large river, the discharge under the bridge affected by backward flow may be estimated by the following equation, in which the corresponding level of large river in converging spot may be adopted as the water level under the bridge:

$$Q_\mathrm{g} = \Omega \frac{\Delta H_\mathrm{g}}{\Delta T} - Q_0 \tag{11.5.2-1}$$

where:

Q_g the discharge under the bridge affected by backward flow (m^3/s);

Q_0 discharge if river at bridge site is at its regular level (m^3/s);
Ω impounding water area upstream of the bridge (m^2), which may be measured from large scale topographic map;
$\frac{\Delta H_g}{\Delta T}$ water rising intensity of the river at bridge site affected by backward flow of large rivers, i.e. impounding rising rate (m/s), which may be substituted with rising rate in higher levels referring to level data from large river observation station.

3. If a design frequency flood event occurs in the river where the bridge is located, and which is affected by backward flow of the design frequency flood in large river but the flood begins to subside, the under bridge discharge may be estimated by the following equations; the water level under the bridge shall be determined through inferring backwater curve according to level of large river in converging spot under design flood frequency. If inferred level is much higher than that of historic backward flow level close to the level of design flood frequency, backward flow level of large river in converging spot and subsiding hydrograph shall be regulated appropriately, the discharge under the bridge recalculated, the under bridge backwater level derived, until it is close to the level of historical backward flow.

$$Q_q = Q'_p + \Omega\frac{\Delta H_t}{\Delta T} \tag{11.5.2-2}$$

$$Q'_p = Q_p\frac{W_p}{W_p - W_a} \tag{11.5.2-3}$$

$$W_a = W_j - W_0 - W_l \tag{11.5.2-4}$$

where:

Q_q the discharge under the bridge (m^3/s);
Q_p design flood discharge of the river at bridge site which is free from the effect of backward flow (m^3/s);
Q'_p increased discharge with the effect of backward flow (m^3/s), $Q'_p = Q_p$ if impounding volume is limited as for non-wide floodplain river upstream of the bridge where landform in both sides is steep;
$\frac{\Delta H_t}{\Delta T}$ impounding decrease rate upstream of the bridge (m/s), which may be determined by referring to subsiding hydrograph of flood level in large river;
W_p design flood volume of the river at the bridge site (m^3), which may be determined by design hydrograph without backward flow;
W_a existing riverbed impounding volume upstream of the bridge (m^3);
W_j impounding volume upstream of the bridge (m^3), which may be determined from topographic map according to the backward flow level under the bridge if flood at design frequency occurs in large river;

W_0 impounding volume below the regular level of channel in reach between the bridge site and the place where regular level of upstream channel equals to backward flow level (m^3);

W_1 water volume above design flood level in reach between the bridge site and the place where upstream design flood level free from the effect of backward flow equals the backward flow level (m^3).

11.5.3 Bridge spans shall be designed in accordance with the following provisions:

1. Discharge shall be calculated according to the third flow condition of Clause 11.5.2, and length of bridge spans shall be calculated according to provisions in Section 7.2 of the *Specifications.*
2. Design elevation of the deck shall be determined according to the provisions in Section of the *Specifications.* As for design level that controls the deck elevation, the backwater level under the bridge shall be derived according to the third flow condition in Clause 11.5.2 of the *Specifications.*

11.5.4 Scour at piers and abutments and embedded depth of foundation calculations shall comply with following provisions:

1. Scour depths at piers shall be calculated according to the three flow conditions of Clause 11.5.2 by referring to the provisions in Sections 8.3 and 8.4 of the *Specifications*, and unfavorable result is adopted as the basis for embedded depth of foundation.
2. As for general scour under the second flow condition in Clause 11.5.2 of the *Specifications*, it shall be calculated under condition of rinsing scour (without riverbed-load movement).
3. Embedded depth of foundation shall comply with the provisions in Section 8.6 of the *Specifications.* Meanwhile, effects of riverbed undercutting by natural scour of large river riverbed shall be taken into account.

11.6 Tidal Reach

11.6.1 The following contents shall be included in hydrological investigation and survey:

1. Investigating tide characteristics in bridge site reach with tidal phenomenon and the degree of tidal effect on the flow.
2. Investigating the historic highest tide flood level in bridge reach and corresponding tidal range, and collecting the data such as measured tide flood level and tide level hydrograph, discharge, tide velocity and velocity hydrograph, storm tide and observation section over the years. If data such as measured level or velocity is unavailable, continuous observation on level shall be conducted in bridge site section and estuary, and observing time shall not be less than one year.

3. Collecting the topographic map of the channel and navigation chart in bridge reach, channel sectional drawing of the vicinity of bridge over the years. If topographic map of the channel is unavailable, it shall be surveyed and mapped according to the *Highway Survey Specifications* (JTG C10); if channel sectional drawing is unavailable, it shall be surveyed and mapped according to the provisions in Clause 5.3.1 of the *Specifications*, and channel evolution and flushing conditions shall be investigated.
4. Collecting the data about design and current situation of hydraulic construction upstream and downstream of the bridge, navigation standard and renovation planning of river reach.
5. If there is freezing and drift-ice situations, data such as ice size, bidirectional movement and up-down float of ice layer shall be investigated and collected.
6. Investigating the height of breakers in the sea and hitting height of wind waves.

11.6.2 Hydrological calculation shall comply with following provisions:

1. If the bridge is located in the upstream of tidal reach, design flood discharge and design velocity may be determined by means applied to reaches free from tidal effects, but tide jacking effect under unfavorable frequency combination in flood season shall be considered for design flood level.
2. If the bridge is located in the downstream of tidal reach, each design value shall be calculated by method of unsteady flow, after determining design tide water level hygrograph based on typical tide pattern selection, and selecting discharge hygrograph under possible unfavorable frequency combination in upstream of tidal reach.
3. If the bridge is located in the midstream of tidal reach, each design value of bridge site shall be calculated according to design frequency of unfavorable combination of flood and tide.
4. If there are no less than 20 years of measured data, such as flow and water level, design flood level and discharge may be inferred according to frequency distribution curve of Pearson III or extreme valueI, and design flood level and discharge hygrograph may be inferred by selecting hygrograph of typical years.
5. If there are measured data from more than 5 consecutive years but less than 20 years, more than 20 years of data is available in nearby survey station, and conditions such as tide nature, influence by river runoff (include flood season), influence of water increase and decrease of both sites are similar, the design flood level may be obtained by using extreme synchronization difference ratio method with the following equation:

$$X_{\mathrm{sy}} = A_{\mathrm{Ny}} + \frac{R_{\mathrm{y}}}{R_{\mathrm{x}}}(X_{\mathrm{sx}} - A_{\mathrm{Nx}}) \qquad (11.6.2\text{-}1)$$

where:

X_{sy} design flood level of the calculation section (m);
X_{sx} design flood level of nearby survey station (m);

A_{Ny}, A_{Nx} annual average sea level at calculation section and at nearby survey station, respectively (m);

R_y, R_x difference between average highest (lowest) tide level and average sea level in the same period of every year at calculation section and at nearby survey station, respectively.

6. If the above Sub-Clauses 4 and 5 are not satisfied, the water level shall be observed continuously with an observation period of not less than 1 year, and high (low) tide level is calculated by the following equations:

$$H_S = H'_S \pm K_k \quad (11.6.2\text{-}2)$$

$$H'_S = A_N \pm (0.6r + K') \quad (11.6.2\text{-}3)$$

where:

H_S design value of high (low) tide level (m);

H'_S tide level value under 10% high tide accumulative frequency or 90% low tide accumulative frequency inferred according to short period observation data;

K_k constant, it may take the value from Table 11.6.2, which has the similar tide property, size of tidal range, river influence and increase or decrease water influence in the nearby harbour;

A_N annual mean height of sea level (m), which may be calculated by referring to short period observation data which is processed on a monthly basis;

r mean tidal range in short period observation data (winter tidal range is not suitable for northern port) (m);

K' constant, 0.4 m may be adopted.

11.6.3 Design of bridge span shall comply with following provisions:

1. In the upstream section of tidal reach, the length of bridge span may be determined using the method for design bridge span in natural state.
2. In the midstream and downstream of tidal reach, the minimum length of bridge span may be determined using the calculated results from Eq. (7.2.1-1) and Eq. (7.2.1-2) in the *Specifications* multiplied by coefficient 1.05–1.15, but this number should not be less than the surface width under mean low tide level over the years.
3. In the vicinity of tidal lock or in reaches affected by reclamation, the changes of channel and hydrological conditions over the years before and after lock construction and reclamation shall be considered for the design of bridge span, especially water level changes affected by reclamation or by local abrupt level up or down if the tidal lock is switched off.
4. If there is freezing conditions or drift ice, the effects of backward and forward flows shall be considered, and the arrangement of bridge spans shall not block

Table 11.6.2 Value of K_k (m)

The location		Water level		The location		Water level	
Sea area	Station	Extreme high water level	Extreme low water level	Sea area	Station	Extreme high water level	Extreme low water level
Huanghai and Bohai	Haiyangdao	0.8	1.4	Donghai	Xize	1.2	1.1
	Dalian	1.0	1.6		Haimen (Zhejiang)	1.4	0.8
	Bayuquan*	1.0	1.3		Dachen*	0.9	1.0
	Yingkou	1.1	1.5		Kanmen	1.6	0.9
	Huludao	1.0	1.5		Longwan (Fujian)	1.4	0.9
	Qinhuangdao	1.0	1.6		Shacheng*	1.1	1.3
	Tanggu	1.6	1.8		Sansha*	1.1	1.3
	Longkou	1.6	1.5		Meihua*	1.0	1.1
	Yantai	1.1	1.2		Mawei	1.4	1.0
	Rushankou	0.9	1.3		Pingtan*	1.3	1.0
	Weihai	1.1	1.1		Chongwu	1.3	1.0
	Qingdao	1.2	1.3		Xiamen	1.5	1.0
	Shijiusuo	1.2	1.2		Dongshan	1.0	0.9
	Lianyungang	1.5	1.2		Shantou	2.3	0.7
	Yanwei	1.1	1.2		Shanwei	1.3	0.7
Donghai	Wusong	1.6	1.0	Nanhai	Chiwan	1.1	1.0
	Gaoqiao*	1.4	1.0		Sishengquan*	1.1	0.7
	Zhongjun	1.3	1.0		Huangpu	1.0	0.7
	Dadaishan	1.0	1.1		Hengmen*	1.3	0.6
	Lvhuashan	1.0	0.9		Denglongshan	1.2	0.6
	Jinshanzui*	1.2	1.4		Dawanshan*	0.9	0.7
	Tanhu*	1.5	1.4		Huangchong*	1.3	1.0
	Zhenhai	1.5	0.9		Huangjin*	1.2	0.8
	Changtu*	1.1	1.0		Sanzao*	1.1	0.8
	Shenjiamen*	0.8	1.0		Zhapo*	1.2	0.8
	Zhanjiang	2.4	0.9		Basuo	0.9	0.8
	Naozhou*	1.3	0.9		Xiangzhou*	1.0	1.1
	Xiuying	1.8	0.7		Shitoubu*	1.1	1.4
	Qinghong*	1.2	0.6		Beihai	1.1	0.9
	Yulin*	0.9	0.6		Bailongwei*	1.3	1.1

Note 1. '*' refer to the value is calculated by conditional joint probability distribution in the station
2. The recurrence interval of K_k is 50 years

the ice drainage. If there are floating big ice-block, anti-pressure or icebreaking measures shall be taken at the piers.

5. Clearance under the bridge and design elevation of the deck shall comply with the provisions in Section 7.4 of the *Specifications*, and the sputter effect on bridge structure by the height of sea breakers shall be taken into account.

11.6.4 General scour and local scour of the bridge may be calculated according to the relevant equations in Chapter 8 of the *Specifications*, and all combinations for design conditions and adverse effects on scour by tide flow shall be considered. This may be determined through hydraulic model test if conditions of hydrology and hydraulics are complex.

11.7 Gulf Regions

11.7.1 Hydrological investigation and survey shall include the following contents:

1. Investigating the characteristics of sea current phenomenon at bridge site area, including flow pattern, velocity, flow direction, flow discharge, tide elevation, wind speed, wind direction, residual current, suspended sediment, etc.
2. Investigating the historical highest tide level, lowest tide level, average high tide level, average low tide level, and the difference among the tide levels at the bridge site area, collecting measured tidal flood levels over the years, flow discharge, tidal flow velocity, storm surge, observed cross-section, etc. If there are no measured water level or discharge data, continuous observation near the bridge site shall be conducted, and observation time should not be less than one year.
3. Investigating the unit weight, salinity and chemical composition of sea water near bridge site area.
4. Investigating the historical sea ice conditions at bridge site area, including sea ice formation time, thickness, compressive strength and scope, the ice size, bidirectional movement and up-down float of ice layer.
5. Investigating the sea breaker height and the height of breakers and waves. If calculating the design wave, annual measured average wave height of maximum wave $H_{1/10}$, wave length, period, wave direction, water depth at the location of buoy for observing wave, sea chart shall be collected.
6. Collecting the underwater topographic map over the years, channel charts, and channel standards at bridge site area. If there is no underwater topographic map, it shall be surveyed according to the requirement of the current *Highway Survey Specifications* (JTG C10), and the sea bottom changes and sediment shall be investigated.
7. Collecting the data on the hydraulic structures design, current situation and renovation plans, etc. near the bridge site.

11.7.2 Design flood level shall comply with the following provisions:

1. Design flood level of bridge in the gulf region may use the high-tide level with recurrence interval of 300 years.
2. The determination of tide level shall be carried out based on the relative height difference between the highest tide level values of years collected at the nearby observation station and each datum level. The Gumbel method (i.e., Type I extreme value distribution law) may be adopted to infer the high tide level value at different recurrence periods.
3. For frequency statistics of design tidal level, fully data of observed tide level for one or more years shall be available. If the data at the newly-built bridge site is observed less than one year, the synchronous correlation analysis may be conducted with the port or tide observation station with data of more than one year, and the short-term synchronization difference ratio method may be used to calculate the design tide level which shall be continuously observed and adjusted later.

11.7.3 Design wave shall comply with the following provisions:

1. The return period of design wave should be 100 years. If measured wave height is bigger than the height at the cumulative frequency of same serial wave in the regulated recurrence interval, the wave height may be calculated according to the measured wave height. If the calculated wave height is larger than the limit wave height in shallow waters, the latter may be adopted.
2. If there are long measured wave data at the bridge site or nearby, the yearly maximum value serial of the wave height at one cumulative frequency in different direction may be used to conduct frequency analysis, and to determine the design wave height of different return periods.
3. Pearson III or extreme value I distribution curve may be used as the frequency curve of wave height, or other theoretical frequency curve may be selected to determine the design wave of different recurrence interval.
4. If there are short measured wave data of one or more years, and the observed big wave data at the bridge site or nearby, the wave height at one cumulative frequency of all observed data without direction may be used as the design wave, and may be determined after comparing and analyzing with the results of other methods.
5. If there is no measured data at the bridge site or nearby, the design wave should adopt the method using wind velocity to calculate wave for gulf and river mouth areas. For open coastal areas, the wave data off the coast should be used to calculate the design wave through shallow deformation.

11.7.4 Design of bridge span shall not only comply with the provision in Clause 4.3.5 of the *Specifications*, but also comply with the relative provisions of the current *Navigation Vessel Bridge Navigation Standards* (JTJ 311). As for the bridge without navigation under it, the design elevation of deck shall not be lower than the sum of the tidal level of design frequency, design wave height, safe value of 0.5 m, and the height of bridge superstructure. As for the bridge with navigation under it, the design elevation of deck shall not be lower than the sum of the design highest navigation level, navigable clearance height, and the height of bridge super structure.

11.7.5 The sediment change caused by seabed natural scour near the bridge site over recent years shall be analyzed according to the observation topographic map. General scour and local scour shall consider the scour induced by flood and tidal current, and also consider the scour induced by wave and tidal current, and select the worst value. This value may also be determined through hydraulic model test with complicated hydrology and hydraulic condition.

Chapter 12
Regulation Works

12.1 General

12.1.1 Necessary regulation works shall be provided to ensure the water drainage and sediment transport through the bridge span, to reduce the adverse deformation of the riverbed and river banks near the bridge site or to resist scour on the subgrade slope.

12.1.2 The general arrangement of regulation works shall be determined according to features of river reaches, natural conditions of hydrology, topography and geology, navigation requirements and situations of hydraulic facilities, as well as by comprehensively considering the effects of high and median low water levels on the deformation of the riverbeds at both banks and the upstream and downstream of the river.

12.1.3 Construction schemes of regulation works shall be designed to consider in line with the design of bridge span. Comparison of several schemes shall be made in terms of technology and economy, and one-sided emphasis shall not be put on large bridge with short dike or short bridge with long dike.

12.1.4 The designed flood frequency of the diversion dike shall be the same as that of the bridge. The designed flood frequency criteria for other types of regulation works may be determined depending on their importance.

12.1.5 The base of a regulating structure located in river channel shall be buried in a depth no less than 1 m below the overall scour line; if it is located at the floodplain, the base shall be buried in a depth no less than 0.5 m below the overall scour line. Plane protection works shall be provided if the requirement for depth cannot be satisfied.

Y. Jiao, *Hydrological Specifications for Survey and Design of Highway Engineering*, JTG Standards and Specifications, https://doi.org/10.1007/978-981-92-3128-7_12

12.2 Arrangement and Scour Calculation of Diversion Dike

12.2.1 For a river channel with single sided floodplain, if the discharge blocked by the bridge approach accounts for 15% of the total design discharge, or for double sided floodplain with the total design discharge divided into two parts by midstream of channel, if the discharge blocked by the bridge approach accounts for 15% of the total design discharge of this side, diversion dike should be provided. If the proportion is smaller than 15%, but the natural average velocity of the blocked flow is greater than 1.0 m/s, a pear-shaped dike should be built. If the proportion is less than 5%, cone slopes of bridge abutments may be strengthened.

12.2.2 An enclosed diversion dike may be arranged near the upstream pass in mountain front alluvial and cross flow reach. In spreading section of middle course, long and large enclosed diversion dikes should not be arranged to constrain the stream. If there are several bridges on a river, peach-shaped diversion dike, water-diversion dike or reinforced embankment may be set between adjacent bridges.

12.2.3 An enclosed diversion dike may be arranged according to the flow and reach conditions if the riverbed is reduced by a bridge span at mountain front changing reach and plain meandering reach.

12.2.4 The plane shape and dimension of diversion dikes shall be drafted through calculations. Necessary adjustments shall be made in accordance with practical experience with operation of diversion dike at upstream or downstream locations, as well as hydrological, topographical, and geological conditions and location of the river reach at the bridge site.

12.2.5 Section of diversion dike should be ladder-shaped and its top width and slope may be adopted according to Table 12.2.5. Special design shall be made if the height of the dike is larger than 12 m or the footing of its slope is submerged in water for a long time.

12.2.6 Elevation of top surface of diversion dike shall be determined by the following provisions:

1. For the enclosed diversion dike, the following equations shall be used:

   Upstream side

Table 12.2.5 Top width and slope of diversion dike

Top width (m)		Slope		
Dike head	Dike body	Dike head	Dike body	
			Dam face	Dam back
3–4	2–3	1:2–1:3	1:1.5–1:2.0	1:1.5–1:1.75

$$H_{ds} = H_s + \Delta Z + \sum \Delta h + L_{ds} I + 0.25 \quad (12.2.6\text{-}1)$$

Downstream side

$$H_{dx} = H_s + \sum \Delta h + 0.25 \quad (12.2.6\text{-}2)$$

where:

- H_{ds} the lowest levee crest elevation of diversion dike at the distance of L_{ds} from upstream of the abutment midline (m);
- H_{dx} the lowest levee crest elevation of diversion dike at downstream of the abutment midline (m);
- ΔZ the maximum backwater height in front of the bridge (m);
- H_s design flood level (m);
- $\sum \Delta h$ sum of factors such as wave run-up, local rush height by tilt water flow, and riverbed surface silting height (m);
- L_{ds} the projected length of the distance between the calculated point of diversion dike and the central line of abutment along the axis of the flow (m);
- I natural gradient of floods in bridge site reach (in decimal term).

2. Upstream side of non-enclosed diversion dikes shall be calculated by the following equations, while the downstream side shall be calculated using the Eq. (12.2.6-2):

If $L_{sh} < L_a$.

$$H_{ds} = H_s + \Delta h_{sh} + \sum \Delta h + 0.25 \quad (12.2.6\text{-}3)$$

If $L_{sh} > L_a$.

$$H_{ds} = H_s + \Delta h \Delta h'_{sh} + \sum \Delta h + 0.25 \quad (12.2.6\text{-}4)$$

where:

- L_{sh} the distance from the point of maximum backwater height on the upstream side of floodplain subgrade to the front edge of abutment (m);
- L_a the distance from the front edge of abutment to the floodplain subgrade of the bank in the same side (m);
- Δh_{sh} the maximum backwater height calculated using Eq. (10.3.3-1) in the *Specifications* (m);
- $\Delta h'_{sh}$ the maximum backwater height calculated using Eq. (10.3.3-6) in the *Specifications* (m).

3. Elevation of each point on the top of pear-shaped dike shall be calculated by the cross slope of the surface, which is calculated using the Eq. (10.3.3-5).

4. If there is drift ice, the crest elevation of levees shall be at least 0.75 m above the maximum drift ice level.

12.2.7 The natural change scour of riverbed and general scour according to Sections 8.2 and 8.3 in the *Specifications* shall be considered for scour calculation of diversion dike, and the local scour of diversion dike itself shall be considered. The maximum scour depth of an existing diversion dike on a similar reach shall also be investigated for verifying the calculated result.

12.3 Arrangement of Spur Dike and Scour Calculation

12.3.1 Arrangement of spur dikes shall comply with the following provisions:

1. Spur dikes shall be arranged according to the guiding regulation alignment and a single long spur dike should not be arranged.
2. A spur dike should not be arranged within twice the length of the bridge from an upstream of the bridge site. Spur dikes may be laid out upstream of the floodplain subgrade to prevent the overbank flow from scouring the embankment.
3. For the watertight spur dike, the projected length perpendicular to flow direction should not exceed 15% of the width of the channel; for spur dikes with drainage property reaching 80%, the proportion should not exceed 25%.
4. Depending on the conditions of the soil and the flow, the dam root may be embedded 3–5 m into the bank, or reinforce the upstream bank by 8–10 m, and downstream by 12–15 m.
5. Crest elevation of non-submerged spur dike may be determined by the provisions in Clause 12.2.6 of the *Specifications*. Crest elevation of submerged spur dike may be determined by the regulating water level, and longitudinal gradient of 0.25–2% should be set on top of the dam. The height of permeable spur shall allow the passage of floating objects from the dam crest.
6. A lifting-water spur dike shall not be arranged on the debris flow ditch.

12.3.2 Dike shape and section form shall be determined by the following provisions:

1. For the scour protection of embankment revetment and foundation of other regulation works, the perpendicular or upturned submerged spur dikes and drowned dam should be adopted.
2. To stabilize channel or to accelerate deposition between spur dikes, the upturned submerged spur dikes should be adopted.
3. During high-flood periods, in order to diverse the flood and protect the bank, subgrade on floodplain and other regulation works from scour, a downturned non-submerged spur dike should be adopted.
4. In the wide and shallow meandering reach with high level of sediments in the flow, a non-submerged permeable spur may be adopted to reduce the effect of oriented flow, decrease velocity and to allow sediments to deposit.

5. Either a flexible structure or a rigid structure may be adopted as the spur dikes. Section shape and dimension of the dikes shall be determined according to the flow conditions and materials of dikes, and section dimension shall satisfy the requirements for stability.

12.3.3 The ross angle between downturned non-submerged spur dike and flow should be 60°–75°; cross angle between upturned submerged spur dike and flow should be 100°–105°; on a convex bank with low flow velocity, the cross angle between spur dike and flow should be 90°. For tidal reach, the spur dike should not be constructed perpendicular to flow at the reach with high velocity on the tide or tidal bore, nearly 15° should be deviated to the perpendicular direction of the upstream; with nearly the same velocity during the flood and ebb, the spur dike may be perpendicular to flow.

12.3.4 Protection length and interval of spur dike shall be determined according to the following provisions:

1. Protection length of spur dike may be calculated according to Appendix E in the *Specifications*.
2. Spur dike interval shall be smaller than the protection length of upstream spur dike.

12.3.5 For the scour of riverbed near the spur dike, the natural change scour of riverbed shall be considered, and the local scour of spur dike itself shall be calculated. The maximum scour depth of existing spur dike on similar reach shall be investigated for verifying the calculated result.

Appendix A
River Reach Classification

See Table A.1.

Y. Jiao, *Hydrological Specifications for Survey and Design of Highway Engineering*,
JTG Standards and Specifications, https://doi.org/10.1007/978-981-92-3128-7

Table A.1 River reach classification

River type	Reach type	Stability		Characteristics of the river and characteristics of riverbed evolution			
		No.	Classification	Morphological characteristics	Characteristics of hydrological silt	Characteristics of riverbed evolution	Major differences between reaches
River in mountain area	Canyon reaches	I	Stable	1. There are many sharp bends on the plane, with wide and narrow riverbeds in the shape of 'V' or 'U' 2. The river profile is mostly in the shape of convexity, with gentle and steep slopes 3. Canyon reaches: narrow riverbed, steep and mostly stony river banks, with no obvious differences between the normal and low flow channel 4. Open reaches: wider river with side floodplains, and sometimes small flood plains and obvious terraces, river islands and sand bars, with gentle slope and fine sand on riverbeds	1. Riverbed slope is generally higher than 0.2% 2. There is large velocity, with the channel's average velocity up to 5–8 m/s during flood period 3. Large water level amplitude, some even reaching about 50 m 4. Small sediment concentration and relatively large bed material particles; due to its big flow and high transportation capacity, if there are river floods, there is pebble movement	1. River is stable, with deformation mostly from slow one-way scour 2. At the entrance or upstream of the narrow mouth of the canyon reach, there is deposition in flood seasons and scour in dry seasons under the influence of damming 3. There is sometimes thick layer of deposit of relatively small particles, and mostly there is scour in flood seasons and deposition in dry seasons 4. The banks play a big role in river constraint and embedding	1. For canyon reaches, the riverbed is narrow and deep, with rocks on the riverbed surface exposed or covered by large boulders. There is steep slope of the riverbed, with large amounts of sharp bends, bayonet, and the section in the shape of 'V' or 'U' 2. For open and straight reaches, the coastline is tidy, the channel stable, and section mostly in the shape of 'U', with distinct floodplains and channels, and flood flows at all levels basically in the same direction
	Open reaches	II III					

(continued)

Table A.1 (continued)

River type	Reach type	Stability		Characteristics of the river and characteristics of riverbed evolution			
		No.	Classification	Morphological characteristics	Characteristics of hydrological silt	Characteristics of riverbed evolution	Major differences between reaches
River in plain area	Straight reaches and slightly bending	II III	Secondary stable	1. Plane shape of rivers in plain areas may be divided into straight and slightly bending, braided, bending, wide floodplain and meandering ones 2. There is wide valley and sometimes the channel is higher than the ground, relying on the dike to converge flow 3. The section of the riverbed is mostly in the shape of wide-shallow rectangle, and usually with distinct floodplain and channel, which is in the shape of oblique triangle at the river bend. The concave side is narrow and deep, wide and there are high point bars along the convex side, with floodplains and sand bars in the transition section 4. With sediment deposits in various forms exposed from the channel in low water season 5. With the large amount of river bends and shallow floodplains in the plain area, there are deep and shallow vertical sections	1. Riverbed slope is gently, generally less than 0.1%, and sometimes less than 0.01% 2. Small velocity, with the average velocity of channel during the flood period being 2–4 m/s 3. Long flood peak duration, and water level and flow amplitude less than those rivers in the mountain area 4. Smaller bed material particles; Sediment transported by water flow is suspended load, mostly sand, silting sand and clay; but there is also riverbed load 5. $\frac{Q_t}{Q_p}$>0.4 or $\frac{Q_t}{Q_c}$>0.67 is wide floodplain river	1. For straight reaches, the normal flow channel is straight and slightly bending with interlocking point bars which move downstream when flooding, and the deep channel on the opposite bank also moves downstream 2. For braided reaches, the channels for normal and high flood are braided, with the trend of periodic alternation of the two branches 3. For bending reaches, there is scour in the concave part and deposition in the convex. For free bending reaches, owing to repeating concave scour and convex deposition, the top of the bending river moves downward and the side of the convex part is twisted and bent downstream in the topography of hogback; at the same time the bending path is lengthened, the resistance increased, neck shortened, and if flooding occurs the bends will become straight 4. Riverbed evolution of wide floodplain winding reaches is similar to that of winding reaches 5. For meandering reaches, the channel is wide and shallow, with large floodplains, which change rapidly, and the mainstream and the branch are constantly changing	1. Difference between stable and secondary stable reaches: channel shoreline, the channels and flood mainstream of the former are basically stable with slow variation; river bend development of the latter moves downward, with its mainstream swinging in the channel. 2. Branches of the braided reaches follow a trend of alternate change. For wide floodplain reaches, overflow covers an area as wide as several kilometers, or even ten kilometers, the width and flow ratios of the floodplain channel are large, the velocity of the floodplain is small, and that of the channel is large
	Braided reaches	III IV					
	Bending reaches	III IV					
	Wide floodplain reaches	III IV					
	Meandering reaches	IV V	Unstable				

(continued)

Table A.1 (continued)

River type	Reach type	Stability		Characteristics of the river and characteristics of riverbed evolution			
		No.	Classification	Morphological characteristics	Characteristics of hydrological silt	Characteristics of riverbed evolution	Major differences between reaches
River before mountains	Reaches with changes before mountains	V	Unstable	1. Reaches with changes before mountains are usually in the area of open space and plains in front of mountains, with the reaches far from the mountain pass, below which are relatively stable plain rivers with many branches, uncertain mainstream movement, unstable shoreline of channels and possible mainstream roll if there are floods 2. Reaches of scour overflow is near the estuary, and the riverbed is steep; because of the monotonous and flat topography, water flow spreads out into a trumpet after flowing out of the pass, leading to sharp decrease of velocity and water depth. And the large amount of sand carried by water flow deposits on the gentle slope in the form of alluvial fan	1. The riverbed slope is between that of the mountains areas and plains, generally 0.1–1%; but that of reaches of scour overflow is sometimes greater than 2–5% 2. The velocity is between that of the mountain area and plain area, and during floods, the average velocity of the channel may reach 3–5 m/s 3. Water flow is wide and shallow; depth amplitude is not large, smaller than that of the mountainous and plain areas 4. Medium or large sediment; in arid and semi-arid regions, it often carries large amounts of fine sediment (both suspended load and riverbed load), which is the main material of deposition	1. For reaches with changes before mountains, characteristics of sediment and riverbed evolution are similar to those of the meandering reaches on the plain, but the slope and sediment particles are larger than the wide floodplain reaches; the characteristics are mainly those of rivers in front of mountains, only with quick and violent course change 2. Reaches of scour overflow usually have no fixed channel, and water flow carrying a large amount of coarse particles leads to either deposition or scour; the mixture of water and sand from steep slope and rushing water flow has great destructive power. After floods, the river branches flow freely, and there are no fixed floodplains which makes it the most unstable reach, the riverbed possibly higher from deposition	1. The differences between unstable and secondary-stable rivers: mainstream of the former swings across the whole river riverbed, with large amplitude and rapid change, and the riverbed may be widened; for the latter, the mainstream swings in the channel with small amplitude. 2. The differences between meandering reaches and reaches with changes before mountains: the former has fine soil particles, deep scour, and fast sedimentation, mainstream swinging not only in the riverbed, but also even leading to river diversion; the latter has coarse particles, shallow scouring, and higher and wider riverbed from deposition and mainstream swinging lead to main channel change, and small widening range 3. Landscapes of reaches of scour overflow generally have the characteristics of the alluvial fan, with its riverbed higher from deposition year by year, more obvious than the meandering reaches, with flood-stream going through the high groove
	Reaches of scour overflow	VI					

(continued)

Table A.1 (continued)

River type	Reach type	Stability		Characteristics of the river and characteristics of riverbed evolution			
		No.	Classification	Morphological characteristics	Characteristics of hydrological silt	Characteristics of riverbed evolution	Major differences between reaches
Estuary	Estuary of delta port	V	Unstable	1. Estuary of delta port is the inland concave-type estuary 2. Estuary of the delta is an alluvial-type extending toward the sea where stand countless point bars and intertwining branches	The slope is generally less than 0.01%, and the velocity is also small; due to tidal effects, the velocity follows a pattern of periodic positive-negative changes; sediment particles are extremely fine, mostly suspended load	The estuary is under the influences of waves and currents. Besides, discharged sediment of the river (after entering the mouth), due to the interaction of currents and runoff, often forms a sand bar, which, combined with sediment particle flocculation from convergence of salty and fresh water, accelerates sediment deposition. In flooding seasons, in reaches controlled by mountains and rivers there may be riverbed scouring, leading to obvious scour and deposition changes on riverbeds of estuary	
	Estuary of the delta	VI					

Note 1. Situations of reaches listed are general circumstances. For example, reach in mountain area is generally stable, but there are exceptions. Some reaches of rivers in mountain area are secondary stable, and even unstable reaches. Under such occasions, the survey and design method shall be decided on the basis of the real situation of specific reaches and its stability analysis.

2. Number in the table refers to the stability of the reach, the smaller the number is, the more stable the reach is; or vice versa

Appendix B
One Dimensional Riverbed Sediment Mathematical Model

B.0.1 One dimensional mathematical model may use one dimensional water flow equation, sediment transport equation, riverbed change equation and relative assistant equation to calculate the riverbed change timely if river encounter the flood, even the process of design flood. The difference between the calculated and initial riverbed elevation is the thickness of riverbed scour. If there is no bridge, the calculated scour is natural scour; if there is a bridge, the calculated scour is local scour.

B.0.2 One dimensional mathematical model shall include the following equations:

1. Water flow equation:

$$\frac{\partial A}{\partial t} + \frac{\partial Q}{\partial x} = 0 \tag{B.0.2-1}$$

2. Water flow motion equation:

$$\frac{\partial}{\partial t}\left(\frac{Q}{A}\right) + \frac{1}{2}\frac{\partial}{\partial x}\left(\frac{Q^2}{A^2}\right) + g\frac{\partial Z}{\partial x} + g\frac{n^2 Q^2}{A^2 R^{\frac{4}{3}}} = 0 \tag{B.0.2-2}$$

3. Suspended load sediment continuity equation:

$$\frac{\partial (QS)}{\partial x} + \frac{\partial (AS)}{\partial t} = -\alpha B \omega (S - S_*) \tag{B.0.2-3}$$

4. Riverbed change equation:

$$\gamma' B \frac{\partial Z_\mathrm{b}}{\partial t} + \frac{\partial (Q_\mathrm{s} + Q_\mathrm{b})}{\partial x} = 0 \tag{B.0.2-4}$$

5. Suspended load transport rate Q_s:

$$Q_\mathrm{s} = QS \tag{B.0.2-5}$$

Y. Jiao, *Hydrological Specifications for Survey and Design of Highway Engineering*, JTG Standards and Specifications, https://doi.org/10.1007/978-981-92-3128-7

6. The bedload sediment transport rate Q_b:

$$Q_b = q_b B \tag{B.0.2-6}$$

where:

B flow cross-section width (m);
A flow cross-section area (m^2);
Q flow discharge (m^3/s);
R hydraulic radius (m);
Z water level (m);
Z_b cross-section average river elevation (m);
S, S_* cross-section average real sediment, flow carrying sediment capacity (kg/m^3);
ω suspended load cross-section average falling velocity (m/s);
γ' sediment dry unit weight (kN/m^3);
g gravitational acceleration (m/s^2);
n roughness;
α sediment saturation recovery coefficient;
q_b bedload sediment transport rate per unit width [kg/(m·s)], selecting the calculate function according to the practical condition;
x spatial coordinate (m);
t time coordinate (s).

B.0.3 Initial and boundary conditions shall comply with the following provisions:

1. Solving the equations in B.0.2, at the upstream entry boundary inputting the known water and sediment process, downstream exit boundary inputting water level process, giving the initial riverbed elevation.
2. When calculation, the water and sediment process may be divided into several parts, the water flow may be to set as the constant in each part, then calculate according to the constant, and simplify the above flow continuous equation and flow motion equation.

B.0.4 The equations may be solved by finite difference method.

B.0.5 One dimensional riverbed sediment mathematical model application shall have the following basic information:

1. Observed riverbed big cross-section or underwater riverbed topographic map.
2. Observed bed material size analyzed results.
3. The latest hydrological analysis and calculation results.
4. Other relative data.

Appendix C
Pier Shape Coefficient and Computing Width of Piers

See Table C.1.

Y. Jiao, *Hydrological Specifications for Survey and Design of Highway Engineering*, JTG Standards and Specifications, https://doi.org/10.1007/978-981-92-3128-7

Table C.1 Pier shape coefficient and computing width of piers

No.	Sketch of pier shape	Pier shape coefficient, K_ξ	Computing width of piers, B_1
1	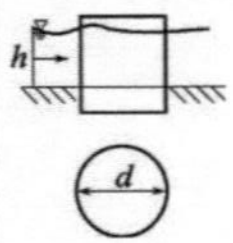	1.00	$B_1 = d$
2	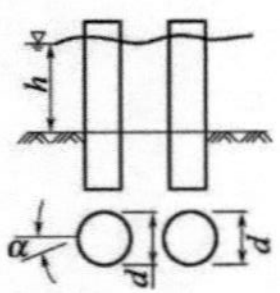	Without connecting beam: $K_\xi = 1.00$ With connecting beam: α: 0°, 15°, 30°, 45° K_ξ: 1.00, 1.05, 1.10, 1.15	$B_1 = d$
3	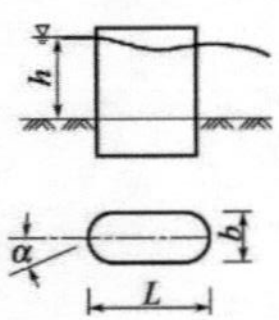	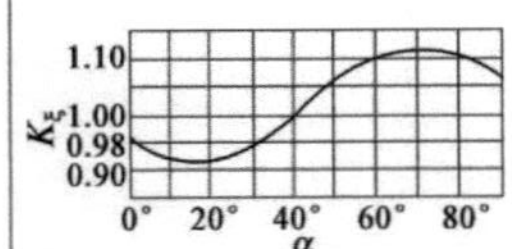	$B_1 = (L - b)\sin\alpha + b$

(continued)

Table C.1 (continued)

No.	Sketch of pier shape	Pier shape coefficient, K_ξ	Computing width of piers, B_1
4	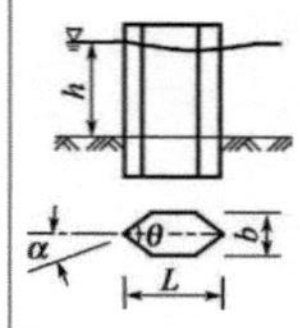	Pier shape coefficient for the cases with various upstream angles if the upstream face is vertical to the water flow: θ = 45°, 60°, 75°, 90°, 120°; K_ξ = 0.70, 0.84, 0.90, 0.95, 1.10 Pier shape coefficient K_ξ if upstream angle θ=90° and upstream face is heterotropic to stream flow 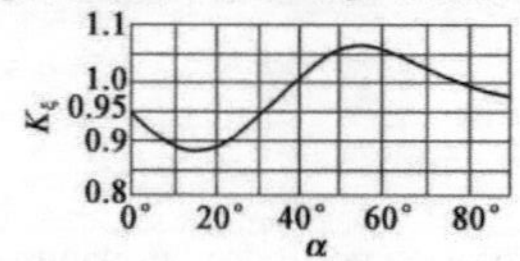	$B_1 = (L - b)\sin\alpha + b$ (Calculated as piers with round ends for simplification)
5	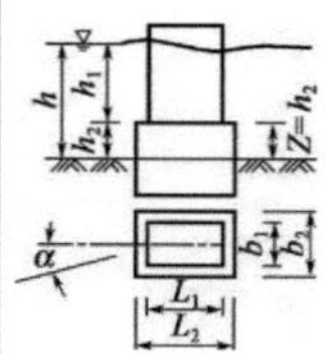	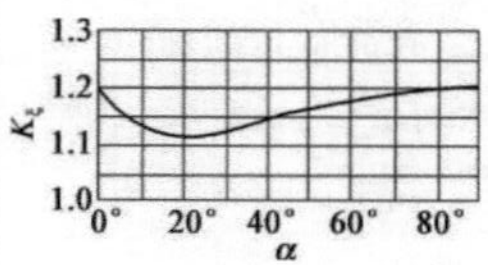	If upstream face is vertical to stream flow $B_1 = \frac{b_1 h_1 + b_2 h_2}{h}$ If upstream face is heterotropic to stream flow $B_1 = \frac{B_1' h_1 + B_2' h_2}{h}$ $B_1' = L_1 \sin\alpha + b_1 \cos\alpha$ $B_2' = L_2 \sin\alpha + b_2 \cos\alpha$

(continued)

Table C.1 (continued)

No.	Sketch of pier shape	Pier shape coefficient, K_ξ	Computing width of piers, B_1
6	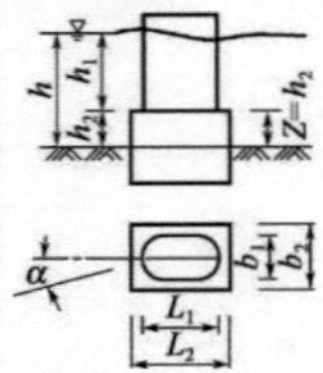	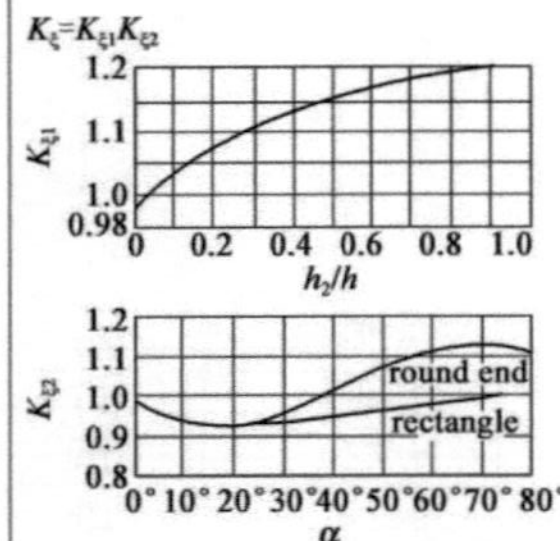 Note: If there is great difference of $K_{\xi 2}$ between open caisson and pier body, it is determined in proportion between the two lines according to the value of $h_1 h_2$	If upstream face is vertical to stream flow $B_1 = \frac{b_1 h_1 + b_2 h_2}{h}$ If upstream face is heterotropic to stream flow $B_1 = \frac{B_1' h_1 + B_2' h_2}{h}$ $B_1' = (L_1 - b_1)\sin\alpha + b_1$ $B_2' = L_1 \sin\alpha + b_2 \cos\alpha$

(continued)

Table C.1 (continued)

No.	Sketch of pier shape	Pier shape coefficient, K_ξ	Computing width of piers, B_1
7	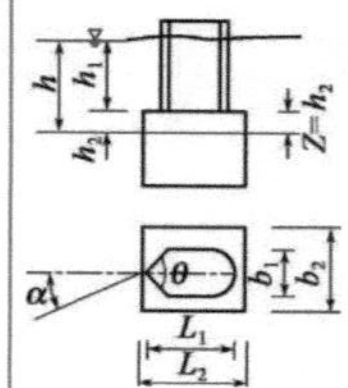	If upstream face is vertical to stream flow $K_\xi = K_{\xi 1}$ 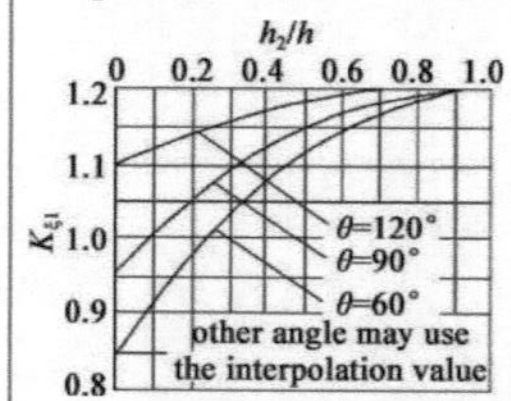 If upstream angle $\theta = 90°$ and upstream face is heterotropic to stream flow $K_\xi = K_{\xi 1} K_{\xi 2}$ 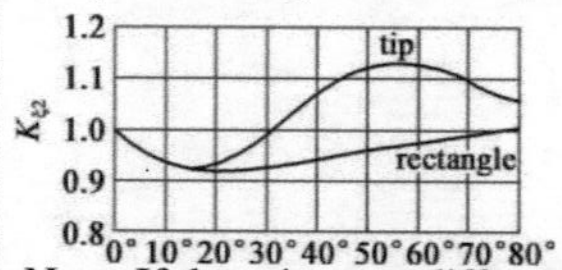 Note: If there is great difference of $K_{\xi 2}$ between open caisson and pier body, it is determined in proportion between the two lines according to the value of $h_1 h_2$	If upstream face is vertical to stream flow $B_1 = \frac{b_1 h_1 + b_2 h_2}{h}$ If upstream face is heterotropic to stream flow $B_1 = \frac{B_1' h_1 + B_2' h_2}{h}$ $B_1' = (L_1 - b_1)\sin\alpha + b_1$ $B_2' = L_2 \sin\alpha + b_2 \cos\alpha$
8	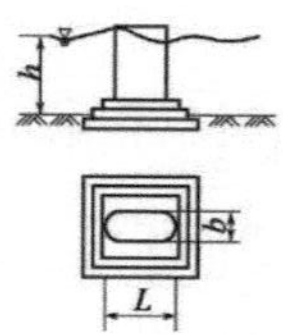	Adopt the pier shape coefficient if upstream face is vertical to stream flow	If upstream face is vertical to stream flow $B_1 = b$ If upstream face is heterotropic to stream flow $B_1 = (L - b)\sin\alpha + b$

(continued)

Table C.1 (continued)

No.	Sketch of pier shape	Pier shape coefficient, K_ξ	Computing width of piers, B_1
9	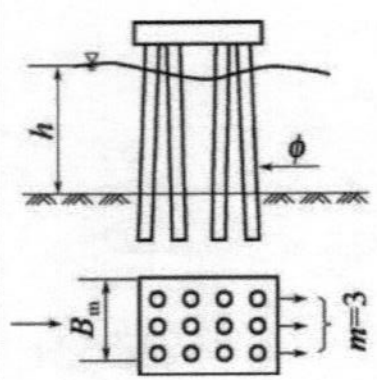	$K_\xi = K'_\xi K_{m\phi}$ K'_ξ—shape coefficient of single pile, determined by pile shape of No. (1), (2), (3), (5) (If most piles are cylindrical, $K'_\xi = 1.0$, which may be omitted); $K_{m\phi}$—coefficient of pile group, $K_{m\phi} = 1 + 5\left[\frac{(m-1)\phi}{B_m}\right]$; B_m—distribution width of pile group vertical to flow direction; m—row number of the piles	$B_1 = \phi$

(continued)

Table C.1 (continued)

No.	Sketch of pier shape	Pier shape coefficient, K_ξ	Computing width of piers, B_1
10	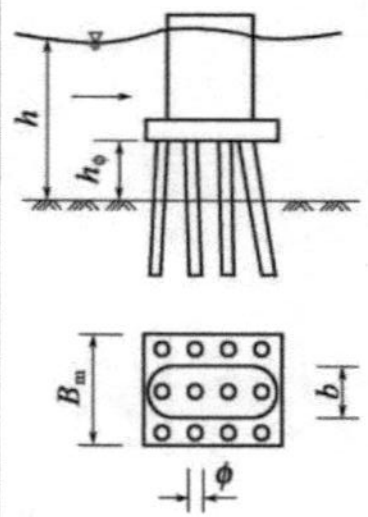	Calculation method for local scour at piers formed by piles and pile cap: If bottom surface of pile cap is below general scour depth, scour depth is calculated according to upper part; if bottom surface of pile cap is above the water surface, it is calculated in the same way as bent frame; if bottom surface of pile cap meets $0 \leq h_\phi/h \leq 1.0$, scour depth h_b is calculated by the following equation: $h_b = (K'_\xi K_{m\phi} K_{h\phi} \phi^{0.6} + 0.85 K_{\xi 1} K_{h2} B_1^{0.6}) K_{\eta 1} (v_0 - v'_0) \times \left(\frac{v - v'_0}{v_0 - v'_0}\right)^{n_1}$ $K_{h\phi}$—reduction factor of submerged cylinder, $K_{h\phi} = 1.0 - \frac{0.001}{(h_\phi/h + 0.1)^3}$; $K_{\xi 1} B_1$—calculated on the condition that bottom surface of pile cap is at general scour depth; K_{h2}—decrease factor of pile body and pile cap; $K_{\eta 1}$, v, v_0, v'_0, n_1 shown in Eq. 65-1; K'_ξ, $K_{m\phi}$ shown in No. (9) 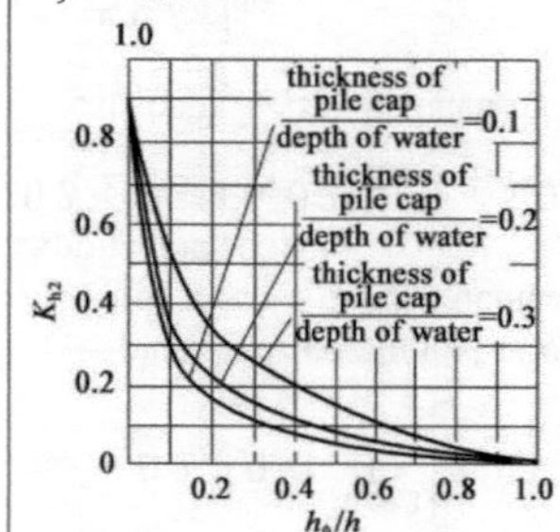	

(continued)

Table C.1 (continued)

No.	Sketch of pier shape	Pier shape coefficient, K_ξ	Computing width of piers, B_1
11	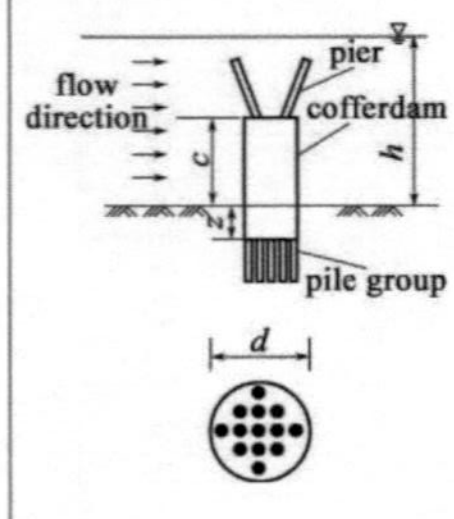	Local scour h_b is calculated by the following equation: $h_b = k_{cd} h_{by}$ $k_{cd} = 0.2 + 0.4\left(\frac{c}{h}\right)^{0.3}\left[1 + \left(\frac{z}{h_{by}}\right)^{0.6}\right]$ k_{cd}—pier shape coefficient of cofferdam pile group with major diameter; h_{by}—local scour depth calculated according to No. (1) pier shape. Applicable scope: $0.2 \le \frac{c}{h} \le 1.0$, $0.2 \le \frac{z}{h_{by}} \le 1.0$	$B_1 = d$
12	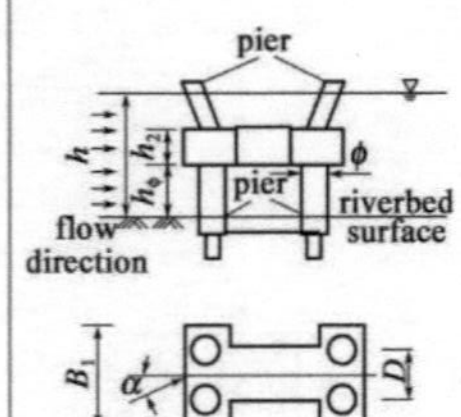	Local scour h_b is calculated by the following equations: $h_b = k_\alpha k_{zh} h_{by}$ $k_{zh} = 1.22 h_{by} k_{h_2}\left(1 + \frac{h_\phi}{h}\right) + 1.18\left(\frac{\phi}{B_1}\right)^{0.6}\frac{h_\phi}{h} h_{by}$—local scour depth calculated $k_a = -0.57\alpha^2 + 0.57\alpha + 1$ according to No. (1) pier shape; k_{zh}—pier shape coefficient of compound pier formed by foundation piles with major diameter and I-shaped pile caps; α—included angle between the axis of bridge and flow direction (radian). Applicable scope: $D = 2\phi$ $0.2 < \frac{h_2}{h} < 0.5, \quad 0 < \frac{h_\phi}{h} < 1.0$ $\alpha = 0 - 0.785$	B_1

Appendix D
Pier Scour with Rock Foundation and Embedded Depth of Basement

See Table D.1.

Y. Jiao, *Hydrological Specifications for Survey and Design of Highway Engineering*, JTG Standards and Specifications, https://doi.org/10.1007/978-981-92-3128-7

Table D.1 Reference data of the pier scouring with rock foundation and embedded depth of basement

Characteristics of rocks					Investigated data		Suggested embedded depth into rocks(classified based on the distance between the average water level of construction in dry seasons and the rock surface) (m)		
Rock type		Ultimate compressive strength (MPa)	Investigated characteristics of bridge rocks with scours		Number of bridge	Maximum scour depth of all bridges (m)	$h < 2$ m	$h = 2$–10 m	$h > 10$ m
			Rock names	Characteristics					
I	Very soft rock	<5	Badly cemented feldspathic sandstone and carbonaceous shale, etc.	The main components are mainly feldspar, and quartz tuff clast and mica come to the second; it is so poorly cemented by clay and iron that hands may shape it into loose sand. Though with marked leaching, it boasts of even rock and fixed fissures. Other rocks such as those with serious airslake and developing fissures, the intensity less than 5 MPa and easily dug with picks and shovels	2	0.65–3.0	3–4	4–5	5–7

(continued)

Table D.1 (continued)

Characteristics of rocks						Investigated data		Suggested embedded depth into rocks(classified based on the distance between the average water level of construction in dry seasons and the rock surface) (m)		
Rock type		Ultimate compressive strength (MPa)		Investigated characteristics of bridge rocks with scours		Number of bridge	Maximum scour depth of all bridges (m)	$h < 2$ m	$h = 2$–10 m	$h > 10$ m
				Rock names	Characteristics					
II	Soft rock	II_1 (Soft rock)	5–15	Clayrock and muddy shale, etc.	The main component is clay, and calcite, chlorite and mica come to the second; it is cemented mainly by clay, and calcium and iron come to the second; there is serious seasoning crack, and is easily air slaked. For rocks under the water, the integrity is good, and is impervious, after exposure, it is easily cracked into pieces, and fragments are rigid, which will collapse into the earthy form when combined with water	10	0.4–2.0	2–3	3–4	4–5

(continued)

Table D.1 (continued)

Characteristics of rocks						Investigated data		Suggested embedded depth into rocks(classified based on the distance between the average water level of construction in dry seasons and the rock surface) (m)		
Rock type		Ultimate compressive strength (MPa)		Investigated characteristics of bridge rocks with scours		Number of bridge	Maximum scour depth of all bridges (m)	$h < 2$ m	$h = 2–10$ m	$h > 10$ m
				Rock names	Characteristics					
		II_2 (less soft rock)	15–30	Sandy shale, interbedded sand-shale, sandstone and conglomerate, etc.	Components of sand-shales are the same as those mentioned above, with coarse particles interbedded; main component of sandstones is quartz, and feldspar and mica come to the second, composed of gravel sand clay. It is cemented mainly by clay, and calcium, with sand coming to the second. There is obvious bedding and joints. Sandy shales are prone to seasoning crack and disintegration at the place where land and water meet	9	0.4–1.25	1–2	2–3	3–4

(continued)

Table D.1 (continued)

Characteristics of rocks					Investigated data		Suggested embedded depth into rocks(classified based on the distance between the average water level of construction in dry seasons and the rock surface) (m)		
Rock type		Ultimate compressive strength (MPa)	Investigated characteristics of bridge rocks with scours		Number of bridge	Maximum scour depth of all bridges (m)	$h < 2$ m	$h = 2$–10 m	$h > 10$ m
			Rock names	Characteristics					
III	Hard rock(less hard rock, much hard rock)	>30	Slate, calcareous sandstone, siliceous rock, limestone, granite, rhyolite and quartzite, etc.	The rock is hard, and the intensity larger than 30 MPa, but if the joints, fissures, and bedding develop fast, the scour shall be taken into consideration. If the rock has complete joints, less fissures and bedding and weak airslake, scour may be ignored. But basement shall be embedded in the rock surface by 0.2–0.5 m	9	0.4–0.7	0.2–1.0	0.2–2.0	0.5–3.0

Note 1. The lower limit of the embedded depth may be adopted if under better conditions; while under poor conditions, the upper limit may be adopted. For bridges under special circumstances, such as those at the lower reaches of dams or those with high water velocity, may be free from the limitation of the data listed in the table

2. In the survey listed, the maximum scour value is got by referring to piers suffering the deepest scour. Suggested embedded depth is calculated and used accordingly for piers at non-mainstream part and that with small velocity, the embedded depth may be appropriately reduced accordingly
3. For the specific name of the rock surveyed in the column of rocks, if used, the rock strength shall serve as the basis of data selection from the table
4. Depth values listed are calculated from the rock surface which includes the weathered layer, and those air slaked into particles or earth are excluded
5. The factor that lithology changes with depth shall be taken into consideration; basement rocks shall serve as the basis and appropriate consideration shall be given to the scour nature of the rocks above the basement
6. Suggested embedded depth listed in the table refers to that of expanding the base or open caisson, and may serve as the location of maximum scour line if pile foundation is used
7. In the columns 'Rock type' and 'Ultimate compressive strength', the rock within racket is regulated by the current specification of rock hard degree

Appendix E
Spur Dike Protection Length Calculation

E.0.1 Non-submerged spur dike protection length calculation shall comply with the following provisions:

1. Straight river, protection length determined by the following equations using flow length after dam, the calculation figure is seen in Fig. E.0.1-1.

   If $(B_c - D_n)/h_c < 70$, i.e., spur dike pick influence the other side

$$L = K_1\left[5.7C_0^{0.3} - \frac{0.09(B_c - D_n)}{h_c}\right]D_n \tag{E.0.1-1}$$

   If $(B_c - D_n)/h_c \geq 70$, i.e., spur dike pick not influence the other side

$$L = K_1(5.7C_0^{0.3} - 6.3)D_n \tag{E.0.1-2}$$

where:

L back flow length calculated from dike head, the straight river is the protection length (m);

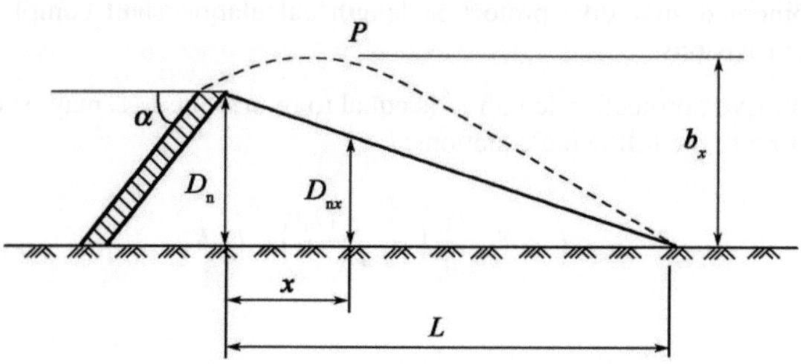

Fig. E.0.1-1 nnCalculation figure for straight river in non-submerged spur dike

Y. Jiao, *Hydrological Specifications for Survey and Design of Highway Engineering*, JTG Standards and Specifications, https://doi.org/10.1007/978-981-92-3128-7

D_n the projection length of spur dike length D at the vertical flow direction (m);
B_c channel width with design water level (m);
h_c channel average depth with design water level (m);
C_0 coefficient, $C_0 = C/\sqrt{g}$, Chezy coefficient $C = \frac{h_c^{\frac{1}{6}}}{n}$, n is roughness, $g = 9.8$ m/s^2;
K_1 correction coefficient of angle α between spur dike and flow direction, $K_1 = \left(\frac{\alpha}{90}\right)^{0.23}$, if $\alpha \geq 90°$, $K_1 = 1.00$.

2. Bend river, protection length L' determined using the relationship of return line and riverside. Return line may be calculated according to the following equations, the calculation figure is seen in Fig. E.0.1-1. The protection length figure of bend river is seen in Fig. E.0.1-2.

$$\frac{b_x - D_{nx}}{D_{nx}} = K_2(5.5 - 7\varepsilon)\left(1 - \frac{x}{L}\right)\left(\frac{x}{L}\right)^{0.8} \quad \text{(E.0.1-3)}$$

where:

L spur dike back flow length (m), calculated according to Eq. (E.0.1-1) or Eq. (E.0.1-2);
ε the same compression ratio, $\varepsilon = \frac{D_n}{B_c}$;
x the distance of any point in return line to the dike head in horizontal (m);
b_x the vertical distance at x point in return line (m);

$$D_{nx} = \left(1 - \frac{x}{L}\right)D_n$$

K_2 the coefficient with the changed between spur dike and flow direction.

$$K_2 = \left(\frac{\alpha}{90}\right)^{0.4} \cdot \left(2 - \frac{\alpha}{90}\right)^{0.1}$$

E.0.2 Submerged spur dike protection length calculation shall comply with the following provisions:

1. Straight river, protection length L' is equal to return line L. L may be calculated according to the following equations:

$$L = 8.3\frac{h_D}{h}\left(\frac{B_c - D_n}{B_c}\right)^{2.6} D_n K_\alpha \quad \text{(E.0.2)}$$

where:

D_n the projection length of submerged spur dike length at the vertical flow direction (m);

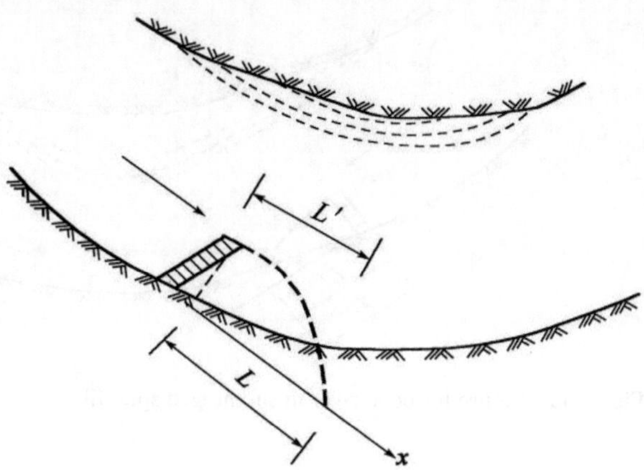

Fig. E.0.1-2 Figure for bend river in non-submerged spur dike

h_D submerged spur dike height (m);
h dike site cross-section average water depth with design water level (m);
B_c water surface width with design water level (m);
K_α correction coefficient of angle between spur dike axis and flow direction.

If $\alpha = 90°$ (perpendicular)

$$K_\alpha = 1$$

If $\alpha > 90°$ (upward)

$$K_\alpha = \left(\frac{\alpha}{90}\right)^{-0.6}$$

If $\alpha < 90°$ (downward)

$$K_\alpha = \left(\frac{\alpha}{90}\right)^{1.3}$$

2. Bend river, if flow direction is to side (bend top upstream), protection length is less than the calculated return line length. If flow direction is off side (bend top downstream), protection length is bigger than the calculated return line length. It may be determined using the Fig. E.0.2 for several conditions.

(1) Flow direction to side, at the dam site plane graph use the dam length D_n as short half shaft, return length L as long half shaft to draw ellipse line which meet the side obtain the point is the end of protection length.
(2) Flow direction off side, at the dam site plane graph use the dam length D_n as a square edge, return length L as the other square edge to draw oblique line and extend to side, the intersection point is the end of protection length.

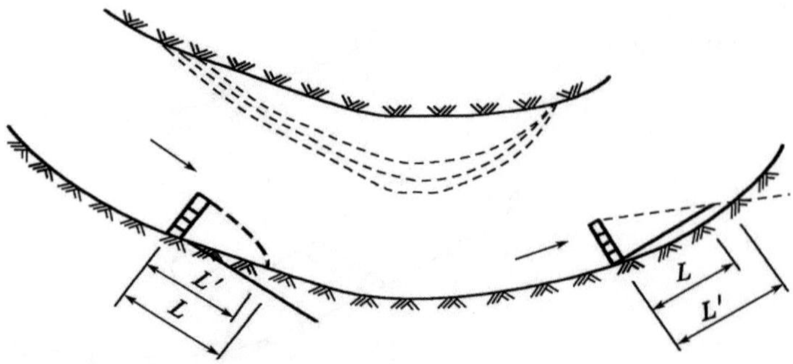

Fig. E.0.2 Figure for bend river in submerged spur dike

Wording Explanation for the *Specifications*

1. Words for strictness in implementation of the *Specifications*:
   (1) 'Must' or 'must not' is used for a mandatory requirement in any circumstances.
   (2) 'Shall' or 'shall not' is used for a mandatory requirement in normal circumstances.
   (3) 'Should' or 'should not' is used for an advisory requirement.
   (4) 'May' or 'may not' is used for a permissive condition that no requirement is intended.
2. The following expression is used in citation of other specifications:
   (1) To state the relation between the *Specifications* and other specifications in Chapter 'General Provisions', it is expressed as 'In addition to the *Specifications*, … shall also comply with the provisions in the current relevant national and industrial standards'.
   (2) When referring to national and industrial standards in clauses of the *Specifications*, it is expressed as 'shall comply with the relevant provisions in ×××××× (×××)'.
   (3) When citing other clauses in the *Specifications*, it is expressed as 'shall comply with the relevant provisions in Chapter × of the *Specifications*', 'shall comply with the relevant provisions in Section ×.× of the *Specifications*', 'shall comply with the relevant provisions in Clause ×.×.× of the *Specifications*', or 'shall be implemented in accordance with the relevant provisions in Clause ×.×.× of the *Specifications*'.

Y. Jiao, *Hydrological Specifications for Survey and Design of Highway Engineering*, JTG Standards and Specifications, https://doi.org/10.1007/978-981-92-3128-7

MIX
Papier aus verantwortungsvollen Quellen
Paper from responsible sources
FSC® C105338

If you have any concerns about our products,
you can contact us on
ProductSafety@springernature.com

In case Publisher is established outside the EU,
the EU authorized representative is:
Springer Nature Customer Service Center GmbH
Europaplatz 3, 69115 Heidelberg, Germany

Printed by Libri Plureos GmbH
in Hamburg, Germany